Building Interactive Entertainment

and E-Commerce Content

e Content

for Microsoft® TV

Peter Krebs
Charlie Kindschi
Julie Hammerquist

PUBLISHED BY
Microsoft Press
A Division of Microsoft Corporation
One Microsoft Way
Redmond, Washington 98052-6399

Library of Congress Cataloging-in-Publication Data
Building Interactive Entertainment and E-Commerce Content for Microsoft TV /
Microsoft Corporation.
    p. cm.
   Includes index.
   ISBN 0-7356-0628-5 ✓
    1. Interactive television.   I. Microsoft Corporation.
  TK6679.3. I58 2000
  384.55--dc21   98-087769

Printed and bound in the United States of America.

1 2 3 4 5 6 7 8 9 WCWC 5 4 3 2 1 0

Distributed in Canada by Penguin Books Canada Limited.

A CIP catalogue record for this book is available from the British Library.

Microsoft Press books are available through booksellers and distributors worldwide. For further information about international editions, contact your local Microsoft Corporation office or contact Microsoft Press International directly at fax (425) 936-7329. Visit our Web site at mspress.microsoft.com.

Acquisitions Editor: Juliana Aldous
Project Editor: Maureen Williams Zimmerman

Table of Contents

Table of Contents

Table of Contents

Table of Contents

Table of Contents

ACKNOWLEDGMENTS

Because the process of producing a book is a group effort, we'd like to thank all those people who helped make this book possible. Thanks go to Barry Potter, Paul Mitchell, Jim Laurel, Phil Goldman, Dan Zigmond, Tim Park, Dean Blackketter, Robert Rouse, Rick Ton, Steve Wilkes, Kiran Rao, Rick Portin, Olinda Turner, Maureen Zimmerman, Juliana Aldous, Katrina Purcell, Marianne Van de Vrede, Murari Narayan, Tom Butcher, Ruston Panabaker, Kilroy Hughes, Tom Cohen, Andy Beers, Alan Yates, Cynthia Thomsen (Hill), Annette Togia, Brian Keller, David Shank, Dana Schmeller, Margot Ayer, Chuck Mount, and Chris Wimmer. Special thanks also go to the editing team at The Write Stuff—Susann Lyon, Jim Hickey, Marilyn Priestley, Shelley Minden, and Roger Sharp—for their hard work.

The author, Peter Krebs, would also like to thank his wife Gloria for her constant kindness, love, and support throughout this project.

The other author, Charlie Kindschi, would like to say ditto to his wife Willow. She steadfastly kept the ship of our life supplied, together, and off the rocks during the stormy schedule we passed through to finish this book. On August 30, 1980, I added another quadriplegic to the world by stuffing my hang glider into a mountain. At the nascent age of five and twenty years I was convinced my life was over. I'd like to thank Dr. Susan Spencer and Dr. Rhinee Yeung, along with the other health care professionals of Group Health Cooperative of Puget Sound's Factoria clinic in Bellevue, Washington, for two decades of unalloyed excellence in keeping what is left of me functional and for reminding me I still have a mind.

The UI designer, Julie Hammerquist, would like to thank her husband, Ken, for his loving support through our trying schedule.

Introduction

Microsoft TV makes the long-awaited dream of interactive TV a reality. This exciting technology opens a world of revenue opportunities for broadcasters, network operators, producers, advertisers, and Web developers. It provides television and Web professionals with a powerful yet flexible platform for developing a limitless range of enhanced TV services, including interactive TV.

With Microsoft TV:

- Broadcasters can offer television viewers a more self-directed experience by providing news headlines, sports scores and statistics, weather, or entertainment information on demand.

- Advertisers can create interactive commercials, enabling viewers to purchase advertised products online.

- Web developers can supply the television industry with a broad range of creative applications that will transform television from a passive appliance into an empowering tool.

Microsoft TV is the customizable software that equipment manufacturers build into television devices such as set-top boxes and integrated TVs. It is part of the Microsoft TV platform, which incorporates the Microsoft TV Server and the Microsoft TV client. Based on Microsoft Windows CE and the Microsoft WebTV Networks experience, Microsoft TV offers a broad range of enhanced communications features, including Broadcast Services (BCS), Microsoft DirectX support, built-in applications, and a Web browser designed specifically for TV.

Microsoft TV supports Web content standards such as HTML 4.0, Cascading Style Sheets level 1 (CSS1), and JavaScript 1.1. This enables content developers to build interactive TV pages using standard Web development tools such as Microsoft FrontPage, Microsoft Visual InterDev, or any other HTML editor. With Web standards and Microsoft TV as a foundation, television and Web development professionals can change the face of television.

Imagine an interactive cooking show that enables viewers to click a button on their remote to see a recipe or to have a copy of the recipe sent to them. Imagine tickers that TV viewers can customize to show selected stock quotes or sports scores. With Microsoft TV, mortgage or insurance companies can collect leads from a commercial or give quotes to prospective clients based on information entered online. Viewers can chat with one another online and use their TVs for voting on civic issues.

Television professionals, network operators, and Web content developers have good reasons for being excited about interactive TV, which is a main feature of Microsoft TV. With the information provided in this book and its companion CD, just about anyone who can create a Web page can design interactive content for Microsoft TV. Whether one is building relatively simple content, such as recipes for a cooking show, or more complex e-commerce applications, the process of creating, delivering, and viewing interactive TV content follows four basic steps:

1. A developer builds interactive TV content, complete with the TV image, using traditional Web development tools such as an HTML editor and a graphics program. The developer then posts the content, as shown in Figure I.1, to a Web server using standard FTP software.

Figure I.I *A Web page created with an HTML editor on the computer.*

2. The television program or commercial that the developer wants to enhance with interactive content can be handed off to a closed captioning service to encode a text string, called an interactive TV link, which specifies the URL of the default page of the interactive content. Figure I.2 shows the video signal with the encoded interactive TV link as it arrives at a TV viewer's set-top box.

Figure I.2 *An interactive TV link to the URL is encoded in the video stream.*

3. When the set-top box receives the video signal with the encoded link, an interactive indicator appears at the bottom of the TV screen, as shown in Figure I.3. The TV viewer clicks a button on the remote control to select the Go Interactive button.

Figure I.3 *The Interactive icon indicates interactive TV content is available.*

4. The interactive TV content, integrated with the TV picture, appears on the screen as shown in Figure I.4. The TV viewer can interact with the content or close it and return to viewing TV in the traditional manner.

Figure I.4 *The viewer clicks the button to interact with the content.*

WHO SHOULD READ *BUILDING INTERACTIVE ENTERTAINMENT AND E-COMMERCE CONTENT FOR MICROSOFT TV?*

Building Interactive Entertainment and E-Commerce Content for Microsoft TV is targeted at television professionals, network operators, advertisers, and Web content developers. However, anyone interested in developing interactive TV content can benefit from reading this book and using its companion CD and Web site.

HOW THIS BOOK IS ORGANIZED

Building Interactive Entertainment and E-Commerce Content for Microsoft TV is broken into parts, some of which may be more relevant to one group of readers than another. For example, a network or cable television executive will want the big picture provided in "Part I: Microsoft TV Primer," but may not want to read "Part II: Microsoft TV Design Guide," which takes a detailed look at design concerns. Similarly, a Web content developer may choose to skim the "Microsoft TV Primer" and focus instead on Web development topics covered in "Part IV: Microsoft TV E-Commerce" and "Part V: Microsoft TV Programmer's Guide."

Part I: Microsoft TV Primer

The "Microsoft TV Primer" is designed particularly for television, cable, and network professionals who want a general understanding of Microsoft TV. It covers the possibilities for creating interactive TV content, explains how Microsoft TV works, and discusses the resources required to create interactive TV content. It also walks developers through the process of creating interactive TV content.

Part II: Microsoft TV Design Guide

The "Microsoft TV Design Guide" is created for Web developers and graphic artists. It describes practical design considerations for showing Web content on TV, including making fonts big enough to be viewed across the room and selecting National Television Standards Committee (NTSC)-safe colors, and it presents advice on designing content with the TV audience in mind. The "Microsoft TV Design Guide" also explains how to layer Web content over a full-screen TV image, how to layer TV over Web content, and how to create a TV object that will display the video signal on the majority of set-top boxes. In addition, the "Microsoft TV Design Guide" offers instructions on using styles and style sheets created especially for TV, adding images and animation, and building Web pages for efficient navigation on Microsoft TV.

Part III: Delivering Microsoft TV Content

This section is intended for broadcasters, network operators, and content providers. It describes the methods supported by Microsoft TV for delivering interactive TV content to the TV viewer's set-top box or integrated TV. Transport methods for delivering interactive TV content include interactive TV links (encoding a text string to a URL in line 21 of the video stream) or multicasting (encoding the actual data and triggers in the video stream). "Delivering Microsoft TV Content" discusses the merits and limitations of interactive TV links and multicasting, shows how to create interactive TV links, and provides a list of vendors that can help you encode interactive TV links into the video signal.

Part IV: Microsoft TV E-Commerce

"Microsoft TV E-Commerce" is designed specifically for Web developers. It covers real-world interactive TV applications that use Active Server Pages (ASP) and Microsoft ActiveX Data Objects (ADO) to write to and read from a Microsoft Access database. Chapter 17, "Creating Forms for Microsoft TV Content," demonstrates how to create a leads-fulfillment form that is tied to a television commercial. Chapters 18 through 20 explain how to build a more complex application that enables TV viewers to purchase a pizza online.

Part V: Microsoft TV Programmer's Guide

The "Microsoft TV Programmer's Guide" consists of a series of brief chapters that interpret the content creation standards espoused by the Advanced Television Enhancement Forum (ATVEF), a cross-industry alliance of companies representing the broadcast and cable networks, television transports, consumer electronics, and personal computer industries. It offers comprehensive coverage of the HTML elements, CSS1 properties, and the Document Object Model supported by Microsoft TV, as well as detailed information about the subset of dynamic HTML (DHTML) supported by Microsoft TV.

BUILDING INTERACTIVE ENTERTAINMENT AND E-COMMERCE CONTENT FOR MICROSOFT TV COMPANION CD

The *Building Interactive Entertainment and E-Commerce Content for Microsoft TV* companion CD, shown in Figure I.5, is a complement to the *Building Interactive Entertainment and E-Commerce Content for Microsoft TV* book, but it can also be used as a stand-alone learning tool. It contains:

- An overview of Microsoft TV and its possibilities for creating interactive TV content

- HTML source code for all the samples covered in the book

- The "Microsoft TV Programmer's Guide," which contains documentation on HTML elements, CSS1 properties, the object model supported by Microsoft TV, and the subset of DHTML supported by Microsoft TV

- A visual gallery of interactive TV content, including samples covered in the book and some samples not covered in the book

BUILDING INTERACTIVE ENTERTAINMENT AND E-COMMERCE CONTENT FOR MICROSOFT TV COMPANION WEB SITE

The *Building Interactive Entertainment and E-Commerce Content for Microsoft TV* companion Web site contains all the interactive TV samples covered in the book. This site should be viewed on a TV connected to a set-top box or on an integrated TV. With the *Building Interactive Entertainment and E-Commerce Content for*

Microsoft TV companion Web site, developers can interact with the samples via their set-top box, keyboard, and remote control—thereby gaining first-hand experience with how interactive television works. The *Building Interactive Entertainment and E-Commerce Content for Microsoft TV* companion Web site, as shown in Figure I.6, is located at *http://www.microsoft.com/tv/itvsamples.*

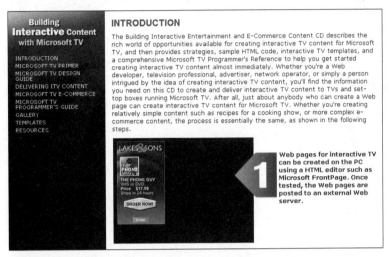

Figure I.5 *The* Building Interactive Entertainment and E-Commerce Content for Microsoft TV *companion CD.*

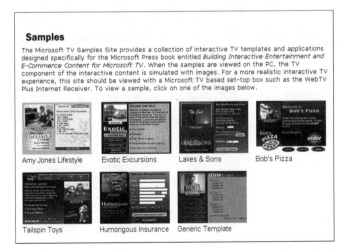

Figure I.6 *Samples Web site at* http://www.microsoft.com/tv/itvsamples.

SUPPORT

Every effort has been made to ensure the accuracy of this book and of the contents of its companion CD and Web site. Microsoft Press provides corrections for books through the World Wide Web at *http://mspress.microsoft.com/mspress/support*. If you have comments, questions, or ideas regarding this book or the companion products, please send them to Microsoft Press using postal mail or e-mail.

Microsoft Press
Attn: *Building Interactive Entertainment and E-Commerce Content for Microsoft TV*
Editor
One Microsoft Way
Redmond, WA 98052-6399
msinput@microsoft.com

Please note that product support is not offered through the above mailing address.

Part I

Microsoft TV Primer

```
//oldimage keep
//oldsrc holds
var oldimage =
var oldsrc = ""

function featureload()
{
        parent.frames["content"].location.href="exotic_feature1.html";
        document.btn_feature.src = featuredon.src;
        if (oldimage != document.btn_feature)
        {
                oldimage.src = oldsrc
        }
        oldimage = document.btn_feature
        oldsrc = featuredoff.src
}

function reservationsload()
{
        parent.frames["content"].location.href="exotic_reservations1.html";
        document.btn_reservations.src = reservationson.src
        if (oldimage != document.btn_reservations)
        {
                oldimage.src = oldsrc
        }
        oldimage = document.btn_reservations
        oldsrc = reservationsoff.src
}
```

Chapter 1

Possibilities for Microsoft TV

In This Chapter

■ Standards Make Interactive TV Content Commercially Viable

■ Types of Content Best Suited for Interactive TV

■ Personalized TV

■ Viewer Participation

Microsoft TV, which provides a software platform for creating interactive TV content as well as other enhanced TV services, will vitalize traditional television programming and attract viewers to television in new, exciting, and engaging ways. Many industry insiders believe that interactive TV promises a virtual gold rush of opportunity.

For starters, Microsoft TV provides a perfect venue for e-commerce, with interactive television commercials leading the way. Some industry insiders predict that interactive TV commercials will be bigger than Internet advertising. According to an August 1999 Forrester Report, interactive television broadcasts will be visible in 24 million households by 2004 and will generate $3.8 billion in commerce. The report predicts that interactive TV will roar through the $5 billion infomercial, home shopping, and direct-response commercial segments and move on to attract consumers with one-click pizza orders, offers to switch long-distance services, and magazine subscriptions.

In addition to e-commerce opportunities, Microsoft TV offers a wide variety of untapped opportunities for creating a better TV experience. For example, with Microsoft TV, broadcasters can provide viewers with a self-directed television experience by offering news, weather, and sports scores on demand, along with customizable stock and news tickers and participation in game shows and opinion polls.

The potential for creating interactive TV content for Microsoft TV is seemingly limitless. In this chapter, we will jump-start your imagination by exploring some of the possibilities for creating engaging, entertaining, and highly profitable interactive TV content. Before we start exploring, however, let's take a moment to discuss how industry standards are making interactive TV not just a technical possibility, but also a commercially viable one.

STANDARDS MAKE INTERACTIVE TV CONTENT COMMERCIALLY VIABLE

Microsoft TV supports existing Web content creation standards, as well as those agreed upon by the Advanced Television Enhancement Forum (ATVEF), including HTML 4.0, Cascading Style Sheets level 1 (CSS1), and JavaScript 1.1. As a result, Web developers can author a single set of interactive TV content that runs reliably not only on Microsoft TV, but also on all devices supporting the ATVEF content creation guidelines. ATVEF's platform-independent standard for adding enhancements to video is the bedrock of interactive TV development. It provides a guideline for a "write once, run on all devices" strategy that extends the reach of interactive TV content to all set-top boxes and integrated TVs built to ATVEF standards. Interactive TV platforms are expected to reach 10 million homes by 2002. Creating content according to ATVEF guidelines provides the widest possible access to this growing audience of interactive TV viewers, thereby offering the best possible return on your interactive TV investment.

TYPES OF CONTENT BEST SUITED FOR INTERACTIVE TV

In a June 1998 Forrester Report entitled "Lazy Interactive TV," authors Josh Bernoff, Christopher Mines, Shar VanBoskirk, and Guy-Frederic Courtin concluded that interactive TV would succeed by embracing "lazy interactivity," a low attention–span paradigm designed for TV viewers. Lazy interactivity, according to the report, involves blending Web content with video in applications that require minimal consumer effort.

What type of interactive TV content is best suited for "lazy interactivity"? Interactive television commercials are a natural fit. Advertisers, working with content providers, can tie interactive content to TV spots, enabling viewers to shop online or request more information with a click of the remote. News, weather, talk shows, and sportscasts are also strong candidates for "lazy interactivity." By contrast, sitcoms and dramas are difficult to enhance with interactive TV content. One TV producer was quoted as saying, "Interactive TV is not an accomplice to narrative storytelling. If I don't want to go to the bathroom during a show, I am not going to go off to the Web."

If you are going to invest time, money, and resources into developing interactive TV content, it is important to develop content that will give you the best return on your investment dollars. The following examples should give you a good idea of the types of content best suited for interactive TV.

Information on Demand

Newscasts, weather, and sportscasts provide good candidates for offering information on demand.

News

With interactive TV, a viewer can get the news she wants, when she wants it. Rather than watching news as presented by a newscast, the viewer can click a button on the remote to browse local, national, or world news headlines; access sports scores; see the four-day weather forecast; or get an update on latest business news, as shown in Figure 1.1.

Figure 1.1 *LKXS News.*

Weather

Interactive weather content can be tied to a 24-hour weather channel or to a news-cast, enabling TV viewers to access the weather forecast at their convenience. With interactive TV, a viewer can review weather maps for more detailed weather infor-mation or check the weather in other areas of the country, an especially handy fea-ture for those planning a trip. See Figure 1.2 for an example.

Figure 1.2 *Island Hopper weather.*

Sports

Interactive TV is bound to please sports buffs because it can simulate what actually happens at a sporting event during slow moments. At some pro baseball games, for example, the stadium scoreboard piques fans' interest between innings by displaying trivia, games, or the scores of other teams playing on the same day. With interactive TV, slow moments during sporting events can be enlivened by these same practices, as shown in Figure 1.3.

Cooking, Gardening, and Home Improvement

Any type of instructional show makes a good candidate for interactivity. Cooking, gardening, and home improvement shows are perfect venues for interactive TV con-tent. In an interactive cooking show, for example, a viewer can click the remote control to see a copy of a recipe or have a copy of the recipe sent to him, as shown in Figure 1.4.

Figure 1.3 *Island Hoppers sports.*

Figure 1.4 *Amy Jones Lifestyle.*

E-Commerce

Television commercials and infomercials provide a natural starting point for e-commerce, but interactive transactions can also be effectively tied into television programming such as concerts or travel shows, as described in the following sections.

Television Commercials

Web designers who build complex sites will find creating interactive TV commercials relatively simple, especially with the help of this book. Designers can build a Web page or pages for a commercial, and then a closed captioning service can encode an interactive TV link to tie the Web page to the TV spot, enabling the TV viewer to order the advertised product with a simple click of the remote control, as shown in Figure 1.5.

Figure 1.5 *Bob's Pizza.*

Interactive TV commercials are not only great for selling products, but also for generating well-qualified leads. For example, in an interactive mortgage or car-insurance commercial, for example, viewers might click a button to request a free quote. Viewers might also enter personal information online and get a quote instantly, or place a request to be called by a representative, as shown in Figure 1.6.

Travel Shows

Travel shows offer ample opportunities for interactive TV content, including interactive TV links to Web pages that offer packaged tours, slide shows of various areas not covered in the program, hotel accommodations, car reservations, and travel tips, as shown in Figure 1.7.

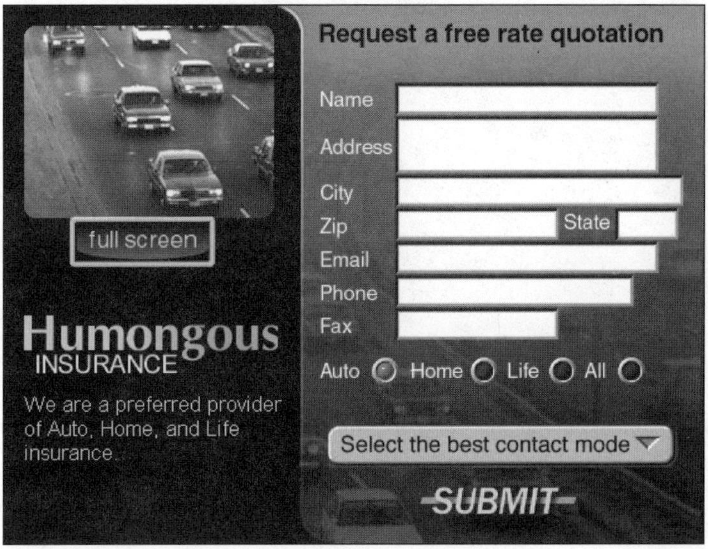

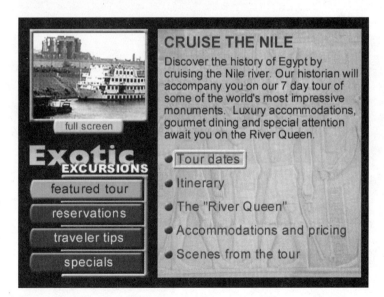

Figure 1.6 *Humongous Insurance.*

Figure 1.7 *Exotic Excursions.*

Concerts, Documentaries, and Television Specials

Television programming such as concerts, documentaries, nature shows, and even movies can be sprinkled with interactive content. A concert, for example, can include an interactive link to enable viewers to buy a CD online. For a documentary or movie, a link can invite the viewer to purchase a DVD or videotape copy of the show, as shown in Figure 1.8.

Figure 1.8 *Lakes & Sons.*

PERSONALIZED TV

With Microsoft TV, content providers can offer viewers a personalized experience. For example, content providers can build a customizable stock or sports updater that enables the viewer to create a ticker showing selected information, as shown in Figure 1.9.

Figure 1.9 *Sports updater.*

VIEWER PARTICIPATION

Game shows and talk shows can retain viewer loyalty by letting viewers participate in such activities as play-at-home games and real-time polling, as shown in Figure 1.10.

Figure 1.10 *Opinion poll.*

WHAT'S NEXT

In this chapter, we have discussed the kinds of TV programming content best suited for interactivity and provided examples to get your imaginative juices flowing. In the next chapter, "Introducing the Microsoft TV Platform," we will discuss how the various components of Microsoft TV fit together to make interactive TV possible. We will also examine how interactive TV content is delivered to the viewer's set-top box and TV screen.

Chapter 2

Introducing the Microsoft TV Platform

In This Chapter

- Microsoft TV Platform Ignites the Tinderbox
- The Genesis of the Microsoft TV Platform
- The Microsoft TV Platform
- Delivering Content to Microsoft TV

In the past, interactive TV technology was more hype than reality. Despite its designation by many industry pundits as the "next big thing," it never quite lived up to its promise. Perhaps the biggest stumbling block was a lack of standards, which resulted in a jumble of proprietary languages and platforms. Without a clear set of standards and a universally accepted platform for developing and delivering content, the industry stagnated.

Enter the Microsoft TV platform, an end-to-end solution for building enhanced TV services and for creating and delivering interactive TV content. The Microsoft TV platform is a modular client-server solution that consists of the following:

- **Microsoft TV**—customizable client software based on the Microsoft Windows CE operating system. Microsoft TV, which original equipment manufacturers (OEMs) build into set-top boxes and integrated TVs, offers such services as e-mail and a Web browser designed specifically for TV.

- **Microsoft TV Server**—a complete and flexible offering of server software based on Microsoft Windows 2000 Server, Microsoft Commerce Server, and Microsoft SQL Server, as well as capabilities new in Microsoft TV Server. With Microsoft TV Server technology, cable providers and Internet service providers (ISPs) can build an entire infrastructure for delivering and managing an interactive TV service.

This chapter describes how the Microsoft TV platform serves as a catalyst for the interactive TV industry. It traces the history of the platform and describes the platform's various components. Finally, it demonstrates how these components work together to create a hotbed of opportunity for cable providers, ISPs, and interactive TV content developers.

MICROSOFT TV PLATFORM IGNITES THE TINDERBOX

For years, the interactive TV industry has been a tinderbox on the verge of catching fire. Now, with the Microsoft TV platform providing a solid foundation on which to create and deliver interactive TV content, the industry is about to ignite, as we describe here.

Standards Pave the Way

The Microsoft TV platform supports industry-wide standards defined by the Advanced Television Enhancement Forum (ATVEF). The ATVEF Specification for Interactive Television 1.1, which is in the process of being adopted by its founders, provides developers with interactive TV content creation and delivery guidelines. This specification ensures that HTML-based content can be delivered reliably to a wide range of set-top boxes and integrated television platforms. It also provides OEMs with guidelines for building TV receivers such as set-top boxes and integrated TVs.

Major Agreements Accelerate Industry-Wide Acceptance

Recent agreements between Microsoft and other industry leaders such as AT&T and Rogers Communications will accelerate the deployment of next-generation broadband and Internet services to millions of homes in North America.

- In May 1999, AT&T and Microsoft announced a series of agreements that call for the companies to work together to accelerate the deployment of next-generation broadband services to millions of American homes. Under the agreement, AT&T committed to use the Windows CE–based system in 5 million set-top devices.

- In July 1999, Rogers Communications and Microsoft announced agreements to deploy and develop advanced broadband services in Canada. As part of the agreement, Rogers chose to license both the Microsoft TV client software and the Microsoft TV Server to support at least 1 million advanced set-top boxes.

TV Networks Are Jumping on Board with Interactive Shows

Television industry interest in interactive TV programming is on the rise, and with the adoption of the Microsoft TV platform by major players in the communications industry, the future of interactive TV looks even brighter. As Josh Bernoff, analyst at Forrester Research, told *Newsday*, "Interactive TV has gone from being a laughing-stock to becoming a potentially massive revenue generator."

For the 1999–2000 season, for example, Microsoft WebTV Networks announced an impressive lineup of television series and specials offering interactive features, including the three leading syndicated series—*Wheel of Fortune, JEOPARDY!,* and *Judge Judy*—as well as *The Today Show, NBC Nightly News with Tom Brokaw, Dateline NBC,* and *The NewsHour with Jim Lehrer.*

As of the time of this writing, NBC, HBO, Columbia TriStar Television, Big Ticket Television, Discovery Channel, Home & Garden Television, MSNBC, MacNeil/Lehrer Productions, E!Entertainment Television, The Learning Channel, and the Weather Channel are offering interactive content on WebTV Networks.

THE GENESIS OF THE MICROSOFT TV PLATFORM

Like many groundbreaking technologies, the Microsoft TV platform can trace its roots to a garage in Palo Alto, California. It was there, in 1995, in a former automobile dealership, that WebTV Networks' three founders—Steve Perlman, Bruce Leak, and Phil Goldman—developed the technology that made WebTV possible.

The founders of WebTV Networks had a common goal of bringing the vast array of information and entertainment found on the Internet into the living room, creating a unique experience that could be shared with family and friends. By developing a product that works with a television set and a phone line—technology found in virtually every U.S. home—the founders achieved their goal of an affordable, easy-to-use product that was a natural extension of the TV-viewing experience.

A year after revolutionizing the industry, WebTV Networks introduced its second-generation product, the WebTV Plus Internet Receiver, which seamlessly integrated television programming and Internet content and services.

Microsoft was quick to recognize this promising new technology. In August 1997, it acquired WebTV Networks for approximately $425 million. Under the purchase agreement, WebTV Networks operates as a wholly-owned subsidiary of Microsoft. Leveraging the experience of WebTV Networks, Microsoft announced in December 1998 that it would provide a set of television-based software platform products and associated network services that were necessary to deliver a complete digital solution for cable operators and broadband service providers.

THE MICROSOFT TV PLATFORM

With its nearly 1 million subscribers, WebTV Networks had proven that there was money to be made delivering interactive TV content. Microsoft, however, had a grander scheme in mind than the then-current WebTV Networks technologies. In late 1999, engineers at both Microsoft and WebTV set to work on the Microsoft TV platform, a client-server software offering based on the Microsoft Windows NT operating system, the Microsoft Commercial Internet System, WebTV Networks technologies, and the Microsoft Windows CE operating system.

As shown in Figure 2.1, the Microsoft TV platform is composed of a client-side software product called Microsoft TV and a server-side software product called Microsoft TV Server.

Microsoft TV Platform

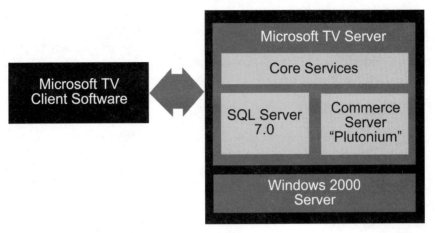

Figure 2.1 *The Microsoft TV platform.*

With the Microsoft TV platform, cable companies, network operators, Internet service providers, or any company with sufficient capital and chutzpa can build a customized interactive TV service similar to WebTV Networks. The steps, which are greatly simplified here, are as follows:

1. Identify the target subscriber market for the interactive TV receiver you plan to offer. Microsoft TV offers three possible configurations that can be built into interactive TV set-top boxes. These range from boxes that offer basic Internet access and e-mail to more advanced boxes that offer chat and interactive TV capabilities.

2. Contract with a manufacturer to build the set-top boxes or integrated TV sets with the chosen configuration.

   ❑ License a copy of Microsoft TV Server and install it on a Windows 2000 server connected to an ISP server farm.

   ❑ Sell subscriptions to your new interactive TV service.

Of course, these steps oversimplify the process of building an interactive TV service company. But the point is, with the introduction of its TV platform software, Microsoft provides an open architecture for enhanced TV services, including building interactive TV hardware, interactive TV services, and interactive TV content, much as the company has provided an open architecture for personal computer devices

and applications. The solid foundation provided by the Microsoft TV platform provides reason to anticipate that a wide variety of new interactive TV devices and services will emerge in the near future.

> **NOTE**   The Microsoft TV platform does not mean the end of WebTV Networks—
> far from it. WebTV Networks service continues to lead the interactive television
> industry as a subsidiary of Microsoft. In fact, a new generation of WebTV Plus
> set-top boxes is being built using Windows CE and Microsoft TV 1.0.

Microsoft TV 1.0

Microsoft TV 1.0 is a customizable software solution for building television devices
such as set-top boxes and integrated TVs. Leveraging the proven benefits of Microsoft
Windows CE and the WebTV Networks experience, Microsoft TV offers enhanced
communication services such as Broadcast Services (BCS), DirectX support, a Web
browser designed for TV, and built-in applications.

Microsoft TV provides several configuration prototypes that OEMs use as the basis
for building TV receivers, ranging from a simple Internet terminal to a fully interactive
digital set-top box. While choosing from the standard configurations, OEMs are free
to differentiate their products by implementing new features, such as DVD players and
recorders, local hard disk storage instead of flash storage, or built-in applications.

How Microsoft TV Works

Figure 2.2 provides an overview of how the components of the Microsoft TV platform work to deliver interactive TV content to the TV viewer. Note that this example
demonstrates only one of two possible methods of interactive TV content delivery.
Both methods are discussed in more detail later in this chapter.

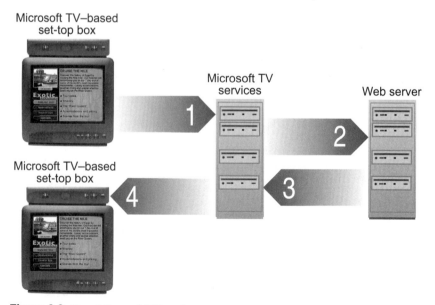

Figure 2.2 *How Microsoft TV works.*

1. A set-top box hosting Microsoft TV software receives an interactive TV link indicating that interactive TV content is available. When the TV viewer clicks the Go Interactive button, a connection to the Internet is established and a request for the Web page defined in the interactive TV link is sent to the Microsoft TV server.

2. The Microsoft TV server calls the Web page from the URL named in the interactive TV link.

3. The Web page is downloaded from the Web server to the Microsoft TV server. If necessary, the Microsoft TV server optimizes the content for TV viewing.

4. The interactive TV content is downloaded to the set-top box and shown on the viewer's TV.

In some cases, the Microsoft TV server optimizes content for TV display. For example, unless font sizes are defined with a Cascading Style Sheets level 1 (CSS1) style, Microsoft TV automatically enlarges fonts, so that text on TV can be read from across the room. In addition, Microsoft TV resizes images wider than 560 pixels to fit within the 560-pixel design space.

> **NOTE** When creating interactive TV content, it is best to design content that will not be altered by the machine running Microsoft TV Server. For more information about how to do this, see Chapter 5, "Guidelines for Designing Microsoft TV Content."

Microsoft TV and Web Standards

Microsoft TV supports interactive TV content creation standards as outlined by the ATVEF, including:

- HTML 4.0
- CSS1
- Level 0 Document Object Model (DOM)
- ECMA 262 Language (ECMAScript)

The Microsoft TV Web browser is designed specifically for displaying TV content. As a result, certain HTML 4.0 elements and properties and CSS1 properties are not supported. By the same token, the Microsoft TV Web browser supports certain TV-centric properties that are not included in the HTML 4.0 or CSS1 subset. To effectively create content for Microsoft TV, it is helpful to become familiar with the content creation standards outlined by the ATVEF. It is also useful to know the limitations and strengths of the Microsoft TV Web browser.

For more information about content creation standards and Microsoft TV's support of HTML 4.0, CSS1, DOM, and ECMAScript, see the "Microsoft TV Programmer's Guide" section on the *Building Interactive Entertainment and E-Commerce Content for Microsoft TV* companion CD.

Microsoft TV Server

The primary customers for the Microsoft TV platform are cable companies, network operators, and ISPs who want to expand into the interactive TV industry. With Microsoft TV Server, ISPs, network operators, and cable companies can build enhanced TV services and manage Microsoft TV–based devices, while leveraging their existing infrastructure and services. In addition, these companies can provide enhanced TV services, including interactive TV, to subscribers using advanced set-top boxes connected to terrestrial, dial-up, satellite, and cable networks. What's more, Microsoft TV Server enables cable companies, network operators, and ISPs to leverage Microsoft Network (MSN) content and services (Hotmail, Search, News, and others) or include their own Internet services.

Microsoft TV Server, based on Microsoft Windows 2000 Server, Microsoft Commerce Server, Microsoft SQL Server, and new Microsoft TV Server technologies, integrates with existing data center systems. For example, Microsoft TV Server can easily plug into existing Internet services, such as e-mail, news, and chat from MSN or other providers. Microsoft TV Server consists of four main services:

- Platform Services
- Device Services
- Application Services
- Deployment and Administration Services

Platform Services

Microsoft TV Server Platform Services provides the underlying foundation to support an operation based on Microsoft TV Server. With Platform Services, network operators can easily integrate their existing network infrastructure, add new components to their service, and create a powerful, flexible system for delivering the next generation of television content. Platform Services includes the following components:

- **Directory System**—a data store for subscriber profiles, user accounts, and other information that is based on Microsoft SQL Server. This system will authenticate Microsoft TV client devices attempting to log on to the server.

- **TV Server Provisioning System (TPS)**—a flexible architecture for provisioning and administering Microsoft TV Server's internal systems. The load generated by service requests such as e-commerce can be spread across any number of servers, even third-party servers. This is very important because it resolves concerns regarding the ability of server systems to handle national or worldwide traffic. Some features of TPS include:

- **Scalability**—each TPS server can scale to accommodate millions of subscriber requests.

- **Robustness**—TPS can be deployed in a cluster of servers with redundant functionality to provide reliable performance.

- **Multiplatform Integration**—TPS provides a cross-platform solution to enable network operators to integrate any back-end system into the TPS framework, effectively leveraging existing systems.

- **Security**—communication between remote components can travel over the Internet or dedicated connections and may contain private information. TPS provides a secure channel for this information.

- **TV Server Security**—enables interconnected systems to exchange information in a secure fashion. This security layer provides vital protection for e-commerce–related transmittals. The system is secure even on cable modem systems that have not implemented the Baseline Privacy Interface (BPI) Specification. Microsoft TV Server has encryption available at 128-bit levels for domestic and authorized use, as well as at 56-bit levels for U.S. export.

- **User Store**—manages server-side user information such as cookies and favorites. This sort of service is vital to support low-end, diskless, or otherwise storage-deprived implementations on Microsoft TV client hardware.

- **Data Warehouse**—a robust logging architecture that captures most Microsoft TV Server internal events, provides complex reporting and analysis capabilities, and is capable of storing large quantities of data. It can also generate user viewing patterns that can be used to create Nielsen-like ratings.

- **Subscriber Management System (SMS) Interface**—connects Microsoft TV Server to third-party SMS systems. This interface enables network operators to integrate their existing subscriber management and billing systems into Microsoft TV Server.

- **Customer Service Representative (CSR) Interface**—enables network operators to extend their customer support infrastructure to include Microsoft TV Server services. In addition, a stand-alone CSR tool ships with Microsoft TV Server.

- **Content Management Environment**—provides an end-to-end framework allowing network operators to author, test, schedule, and deploy content to Microsoft TV Server, using any content development tool.

- **Content and Ad Targeting Service**—enables network operators to target content and ads according to specific user profiles by leveraging the power of Microsoft Commerce Server. Advertisers can benefit from this service's detailed information on user zip code, age group, and recently visited Web sites.

- **Content Reformatting and Proxy Server**—caches Internet content for improved performance, transcodes media into formats optimized for display on television, and reformats Web pages to improve performance and display. The Content Reformatting and Proxy Server extends and improves the morphing done on Microsoft TV service servers. Content Reformatting Modules (CRM) expand the functionality of Internet Information Services (IIS) by adding modules that map streaming media, such as audio files, for play in the set-top box space. The Proxy Server may also be used like a V-chip to create walled gardens of approved content for minors.

Device Services

Device Services enable Microsoft TV Server to administer, configure, and update client hardware from a remote location. These services can quickly diagnose and fix many customer support issues. With automatic operating system upgrades, software updates, and application downloads, the network operator can provide a reliable, efficient service to end users. Software updates and fixes are simply sent to a Microsoft TV client through its back channel modem. Device Services include the following components:

- **Authentication and Connectivity Services**—enables set-top boxes to be authenticated on a network. These services can also authenticate and provision a cable modem.

- **Client Configuration Server**—allows network operators to configure client installations, download new client software and/or operating system upgrades, and upload data from the advanced set-top box to Microsoft TV Server.

- **Electronic Program Guide (EPG) Server**—aggregates program listing data from multiple sources and transmits it to client hardware via a format compatible with the Microsoft TV client's EPG application. This server, which allows users to search by program name or type, is at the top of favorite feature lists compiled by set-top box owners.

Application Services

Through its Application Services, Microsoft TV Server provides a core set of applications, including e-mail, home page, and content centers. Application Services also include a Client Simulator, which assists network operators in creating their own value-added content for end users. Application Services include the following components:

- **Web Applications**—these include an application for creating home pages, called Web-Home, as well as content centers, Web search, and Internet e-mail. The various Web Applications may each be customized to give brand name and service identity to a network operator service.

- **Client Simulator**—enables developers to build and test applications in an environment that emulates Microsoft TV.

Web-Home, the first page the user sees upon connecting to the Internet, provides a starting point for the user's Microsoft TV experience. Network operators can customize this page to include their own branding; links to their own content; and links to Web-based services, advertisements, and promotions, as shown in Figure 2.3.

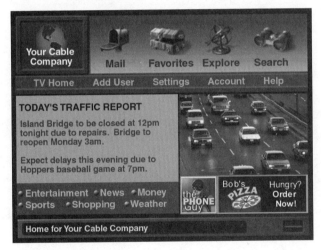

Figure 2.3 *Customized home page for a service provider.*

Deployment and Administration Services

Microsoft TV Server's robust administration and management services enable network operators to run an efficient, reliable service. By incorporating existing third-party network management software, network operators can simply plug the TV Server infrastructure into the existing management operations. With the added benefit of Microsoft Management Console snap-ins, a network operator can administer TV Server and all its related external services from an easy-to-learn interface.

- **Simple Network Management Protocol (SNMP) interface**—enables network operators to use third-party system management tools to manage Microsoft TV Server, providing powerful administrative capabilities.

- **Microsoft Management Console (MMC) snap-ins**—allow Microsoft TV Server's administrative capabilities to share a common look and feel. These components are compatible with SQL Server and related management services.

- **Setup and Configuration Service**—allows network operators to set up automatic installations, configure existing installations, and apply new installations of Microsoft TV Server.

DELIVERING CONTENT TO MICROSOFT TV

In September 1999, the ATVEF formally announced that a royalty-free license for its enhanced content specification was available to its more than 75 worldwide founders and adopters. The specification produced by the ATVEF lists interactive TV content creation and delivery standards. This specification helps interactive TV developers create and deliver compelling content that will work on a variety of TV receivers. The methods defined by the ATVEF are described in more detail in the following sections.

Understanding Front and Back Channels

Most TV receivers built to ATVEF guidelines have a coaxial cable that carries a unidirectional (read-only) video signal and a second cable or phone line that is connected to a modem. For clarity, we will call the coaxial video cable the *front channel,* and the phone line or coaxial cable connected to a modem, the *back channel*.

Front Channel

There are several ways that a TV signal can be delivered via the front channel, including:

- coaxial cable
- satellite
- terrestrial (transmission through the air)

Cable, satellite, and terrestrial video broadcasts may be either analog or digital, but they are always unidirectional, with information flowing from the broadcaster to the receiver. Currently, most video is transmitted in analog format. Receivers complying with the specification adopted by the ATVEF must be able to receive and decode Transmission Control Protocol/Internet Protocol (TCP/IP) data packets, as well as video over their front channel.

Back Channel

The back channel on a TV receiver conforming to the specification adopted by the ATVEF may be a standard personal computer modem, as is used in WebTV Plus set-top boxes, or a cable modem, as used by the new Motorola and Digital Instrument set-top boxes. Another new back channel solution is Digital Subscriber Line technology (DSL). However, at the time of this writing no TV receivers contain DSL capability.

The back channel provides the capability for two-way communication, enabling TV viewers to interact with content. For example, with a modem connection to the Internet, the viewer can purchase products, submit requests for more information, vote, or chat online.

> **NOTE** Microsoft TV client software includes drivers for broadband back channels, such as cable and DSL modems. Speed and a continuous connection are two of the advantages of DSL and cable modems over the standard personal computer variety of modems.

Transports A and B and the Vertical Blanking Interval (VBI)

The specification prepared by the ATVEF defines two methods for delivering interactive TV content, called Transport A and Transport B. Because two wires enter the TV receiver (front channel and back channel) and there are two methods of transport, it is easy to make the mistake of thinking that Transport A corresponds to the coaxial video feed and Transport B corresponds to the back channel modem. Actually, Transport A is distinguished from Transport B by the way interactive TV content is packaged and which wire carries it.

Transport A uses text strings called triggers and links sent over part of the front channel. Then Web pages and their supporting images are brought in using the back channel. Transport B uses TCP/IP packets to send both Web data and triggers on the front channel.

Transport A

Transport A is by far the most popular mode of transmitting interactive TV content in the analog video space. In a Transport A scenario, the URL address to a Web page is sent via the video cable (front channel) as an ATVEF trigger (link). This trigger appears on-screen as a prompt to tell the viewer that interactive TV content is available. When the prompt appears, the user initiates an Internet (back channel) connection by pressing a button on an infrared remote control or wireless keyboard. After a back channel connection is established, the TV receiver downloads the Web page addressed in the trigger. After the first page arrives, Transport A script triggers may be sent that fire functions shaping the user's interactive TV experience. Most of the load (data) in a Transport A scenario comes in over the back channel.

Transport B

Transport B gets interactive TV data to the user by packing up Web pages and their supporting images as TCP/IP packets and sending them down the line using a unidirectional protocol called UHTTP and a reliability layer called UDP. The acronyms UDP and UHTTP stand for User Datagram Protocol and Unidirectional Hypertext Transfer Protocol. Announcements and triggers are also sent as TCP/IP packets in Transport B. Typically, Transport B happens on a digital coaxial cable that is also used to bring in video signal (front channel). The back channel in Transport B is used only for two-way communication, like the submission of a form. There is no doubt that if bandwidth in the video signal were free and plentiful, Transport B would be more popular. However, at this time, limited bandwidth restricts the use of Transport B.

Vertical Blanking Interval (VBI)

The vertical blanking interval (VBI) provides a way for digital information to be transported in an analog signal. Each time a television picture redraws, there are ten blank lines at the top of the screen available for the transport of data. Because National Television Standards Committee (NTSC) television pictures interlace, it takes two screen redraws to make each new television picture. Each one of the interlaced redraws carries effectively 21 VBI lines with which to work, each roughly equivalent to 9,600 baud on a standard computer modem. Transport A triggers are sent over a special part of the VBI called Text 2, Line 21. Incidentally, Line 21 of the VBI is the line used to carry closed captioning information for those with hearing impairments. Transport B data can be sent over several lines of VBI, but VBI lines in the television industry are precious commodities. Transport B is largely a digital TV tool.

Figure 2.4 shows how triggers and data are encoded into the VBI and delivered to the TV viewer. With the Transport A method, the trigger is encoded into the VBI, and interactive TV data is delivered via the back channel. With the Transport B method, both the data and triggers are encoded into the VBI or a digital signal and delivered via the front channel.

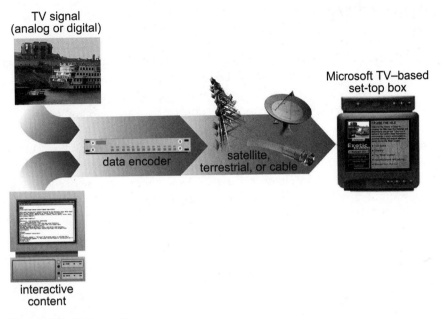

Figure 2.4 *VBI encoding.*

Digital Video Signals

Although the combination of video encoders and the VBI provides a good way to deliver content in today's predominantly analog world, digital transmission is also possible. One current example of digital delivery is the MPEG2 stream sent to satellite set-top systems like EchoStar. Expect to see digital transmission of television eclipse analog transmission in the near future. In a digital world Transport B will likely be the norm, and Transport A will slip into obscurity.

FOR MORE INFORMATION

■ For more information about WebTV Networks, see *http://www.webtv.net/*.

■ For more information about the ATVEF, see *http://www.atvef.com/*.

■ For general information about the Microsoft TV platform and Microsoft's TV solutions, see *http://www.microsoft.com/tv/default.asp*.

WHAT'S NEXT

Interactive television promises enormous business opportunities for network operators, OEMs, content providers, broadcasters, application developers, and others in the television field. Because of the complexity of developing and delivering hardware and services, it is wise to leverage proven standards and products to their best advantage. Microsoft TV Server, based on proven Microsoft Windows and Microsoft BackOffice technologies, meets the primary challenges of this market by providing great functionality, cost effectiveness, ease of use, and reliability. This chapter introduced the Microsoft TV platform and discussed how the pieces fit together to enable you to create and deliver interactive TV content. The next chapter describes the equipment you need to create and deliver interactive TV content for Microsoft TV.

Chapter 3

What You Need to Create and Deliver Microsoft TV Content

In This Chapter

■ Minimum Setup for Creating Microsoft TV Content

■ Recommended Setup for Creating Microsoft TV Content

■ Setup for the WebTV Plus Service

■ Setup for the WebTV Viewer

■ Setup for PWS

■ The Ultimate Interactive TV Studio

You may be surprised to discover that you already have most of the tools necessary to create content for Microsoft TV. In fact, you can get started creating interactive TV content with the standard Web tools on most computers. Most likely, however, you will want to build a development environment that is optimized for creating interactive TV content.

This chapter describes a bare minimum setup necessary to get started creating interactive TV content, outlines a recommended setup, and then offers a glimpse of the ultimate interactive TV studio.

MINIMUM SETUP FOR CREATING MICROSOFT TV CONTENT

To get started creating Microsoft TV content, all you really need is a personal computer, an HTML editor, a graphics program, and a simulator called the WebTV Viewer. The minimum setup also works well for graphic artists who design Web pages but are not necessarily responsible for coding and testing them.

Hardware Recommendations

You can create Microsoft TV content on a Windows-based or Macintosh personal computer. For the Windows-based computer, the following features are recommended:

- 166-megahertz (MHz) CPU or faster
- 32 megabytes (MB) of RAM or more
- CD-ROM drive
- 2-gigabyte (GB) hard disk drive
- A 56-kilobits per second (Kbps) modem or an Ethernet connection to a local area network (LAN) that provides you with Internet connectivity

For design purposes, a color printer is also handy but is not required. Also, graphic artists may want to attach a computer-to-video scan converter and a TV to their personal computer to check the output of their designs on a television set.

Software Recommendations

As a basis for creating content for Microsoft TV, the following software is recommended.

Operating System

You will need a computer running Windows 95, Windows 98, Windows 2000, or Windows NT Workstation 4.0 or later.

HTML Editor

You do not need an expensive HTML editor to get started developing interactive TV content. Notepad, the free text editor that comes with Microsoft Windows, works just fine for creating HTML pages. You can also use Microsoft Visual InterDev, Microsoft FrontPage 97 or later, or any other third-party HTML editor that you are comfortable using.

Graphics Software

The industry standard for professional graphics software is Photoshop from Adobe. Photoshop is great for creating content for TV because it includes a National Television Standards Committee (NTSC) filter for creating NTSC-safe colors. In general, however, you can use any graphics program that you traditionally use for creating images for Web pages.

WebTV Viewer

WebTV Networks provides a valuable design tool called the WebTV Viewer 2.0 that simulates Microsoft TV on the personal computer. At the time of this writing, you can download the WebTV Viewer from *http://developer.webtv.net/design/tools/viewer/*.

For instructions on setting up the WebTV Viewer tool, see "Setup for the WebTV Plus Service" later in this chapter.

RECOMMENDED SETUP FOR CREATING MICROSOFT TV CONTENT

The hardware and software previously described in the "Minimum Setup for Creating Microsoft TV Content" section gets you started quickly, but you will most likely want to actually view your content on a set-top box connected to a TV. To view your content on TV, you need to add a few components to the minimum setup, as shown in Figure 3.1.

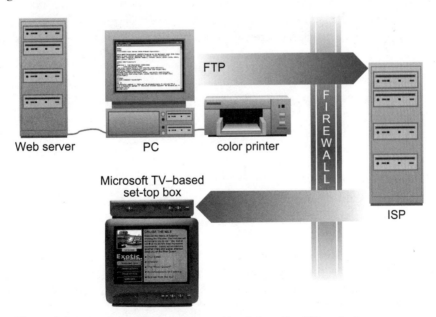

Figure 3.1 *Recommended setup for creating interactive TV content.*

Hardware Recommendations

The following sections list the hardware recommendations for creating interactive TV content for Microsoft TV.

Personal Computers

To design content for Microsoft TV, we recommend that you use two personal computers:

- One reasonably modern personal computer with 166-MHz CPU or faster, 32 MBs of RAM or more, CD-ROM drive, 2-GB hard drive, 56-Kbps modem or an Ethernet connection to a LAN that provides Internet connectivity.

- A second slightly more powerful Web server, including a 233-MHz or faster CPU, 64 MBs or more of RAM, an 8-GB hard drive, and Ethernet cards for both the server and the development computer. Depending on your LAN, you may need a personal hub to connect both machines. If you are unsure, contact your network administrator or computer equipment retailer.

WebTV Internet Receiver and a Color Television Set

The WebTV Internet Receiver is a set-top box that connects to a TV set and a phone line. You can purchase this type of receiver at electronic retailers. The cost varies, but a WebTV Internet Receiver can generally be purchased for about $200. This includes a remote control, audio/video cable, 25 feet of phone line, and an RJ-11 phone line splitter. An optional wireless keyboard, which is highly recommended, is available for about $75.

Local Area Network

You can develop interactive TV content on a stand-alone computer, but for the serious developer, a LAN is a necessity. You can create a LAN using a second machine. The second machine may be a Windows NT server running Internet Information Server (IIS), a Windows 2000 server, or simply a Windows 98 machine running Personal Web Server (PWS). It is particularly important to have a separate Web server when it comes to staging e-commerce applications. You will want to make modifications to your back-end database as you go along. It is not realistic to transmit large database files via File Transfer Protocol (FTP) very often. Besides, if something is wrong with the Digital Switched Network (DSN) and no data is being written to the data store, you are going to want to verify that everything works on your end before you call your Internet Service Provider (ISP). To test DSN functionality, it is best to have a second computer connected via Ethernet LAN running Windows 98 and PWS, Windows NT and IIS, or Windows 2000.

> **NOTE** By the time this book is published, there may be receivers other than WebTV that contain the Microsoft TV client. Any set-top box or integrated television that is using Microsoft Windows CE as its operating system will contain the Microsoft TV client.

We will discuss setup and use of the WebTV receiver in a moment. For now, here are the rest of the requirements for the recommended Microsoft TV content creation setup.

Software Requirements

To create content for Microsoft TV, the following software is recommended.

Operating Systems

If you want to emulate the equipment on the server of your ISP, we recommend that you get Microsoft Windows NT with IIS or Microsoft Windows 2000 Server. This one is your call. You can also use Windows 98 with PWS, as explained later in this chapter.

IIS is part of Option Pack 4 for Windows NT 4.0. To get IIS, check the Microsoft Web site or your computer retailer.

HTML Editor Software

If you are going to get into interactive TV content development in a serious way, you need a serious Web development platform. In the Web development arena, Microsoft Visual InterDev is the top of the line and works well for creating robust e-commerce content.

Database Software

Microsoft Access is a great tool for many database applications. Because of its relative ease of use and ubiquitous deployment, Access is used in this book's examples. Unfortunately, Access does not scale for very large e-commerce projects. As a general rule, if you expect to have more than 60 simultaneous users reading or writing data, you should use Microsoft SQL Server 7.0. You can get a single-license copy of SQL Server 7.0 in Microsoft Office 2000 or Microsoft Visual Interdev.

For information on SQL Server site licenses or other Microsoft BackOffice products, visit *http://www.microsoft.com/backoffice/*.

Internet Service Provider

To view interactive TV content over a Microsoft TV–based set-top, you need to view the content on an external Web server. That means you need an ISP. Some ISPs provide only Unix Web servers. While a Unix server works for hosting interactive TV content, it will not run the Active Server Pages (ASP) content covered in this book. For best results with this book, make sure the ISP has Windows NT or Windows 2000 Server to host your test pages.

All but the smallest ISPs have economical plans that include e-mail and Web site hosting. Fees range from $20 to $40 a month. Generally, an ISP fee is tied to the amount of hard disk space reserved for you on a Web server. In the beginning, 20–30 MBs of hard disk space will be adequate for your needs. Later, when you land that big account, you may need a whole server of your own. Ask potential ISP firms if they offer Web server hosting and maintenance.

When you are shopping for an ISP, try to locate one that allows Data Source Name (DSN) hosting. DSN hosting is the way Web pages write information to data management systems like Microsoft Access or SQL Server. It is normal to pay an additional monthly fee for DSN service. You will learn all you need to know about setting up Open Database Connectivity (ODBC) and DSNs in Chapter 18, "Creating Bob's Pizza."

FTP Software

File Transfer Protocol (FTP) software is used to get files from a development machine to a Web server hosted at an ISP. If you buy a shrink-wrapped FTP software program, you'll probably be the first one who ever did. There are scores of free or shareware FTP programs available. Check your ISP support Web site for directions on setting up your FTP software. Often ISPs will recommend a particular FTP program to their subscribers and provide setup instructions that are based on that program. Two very popular shareware FTP utilities are Voyager and Cute-FTP.

To find an FTP program, check out Shareware.com, the Web site maintained by CNET at *http://www.shareware.com/*.

> **NOTE** Shareware is not free. After the trial period you should send a check to the manufacturer of whatever FTP program you settle on.

The WebTV Plus Service

In addition to the ISP you use for normal Internet connectivity, you also need the Microsoft WebTV Plus service and a WebTV-based Internet receiver. These are required to view and test interactive TV content over a set-top box. The procedure for setting up the WebTV Plus service and a WebTV-based Internet receiver is described in the following section.

SETUP FOR THE WEBTV PLUS SERVICE

To set up the WebTV Plus service, you need the following equipment and services:

- **WebTV Plus Internet Receiver with keyboard**—Considered optional, they are, in reality, required.

- **Color television.**

■ **Cable TV service**—Part of the hookup for a WebTV Plus Receiver is a video feed. You will see more about cable service in the section about setting up and using WebTV.

■ **Analog phone line**—Many offices are set up with digital phone line systems. In order to connect the modem on the back of a WebTV set-top box, you need an analog line. If you are unsure what kind of line you have, check with your office manager or telecommunications support staff.

■ **Internet Service Provider (ISP)**—WebTV Plus offers the option of using your own ISP for accessing WebTV. However, many people opt to simply use the WebTV Plus service, which automatically assigns an ISP that works in concert with the WebTV Plus service to deliver Internet content to the WebTV-based Internet receiver. When you use WebTV Networks as your main connection, you get a monthly bill, currently $24.95. When you use your own ISP to access WebTV, you get two bills. One bill, for $9.95 a month, is from WebTV Networks. The second bill is from your ISP.

NOTE The WebTV-based Internet receiver connects to a standard analog phone line. Many business lines are digital, so it may be necessary to bring a new, separate line into your office or work area. If you are not sure what kind of line you have, ask your technical support or telecommunications personnel.

Setting Up the WebTV Plus Receiver

To set up the WebTV Plus Receiver, follow the instructions in the setup manual that comes with the receiver, along with the online instructions supplied by WebTV Plus.

Viewing Web Pages with the WebTV Plus Receiver

The WebTV Plus service provides a home page that automatically organizes content on the Web for you. In addition, the WebTV Plus service provides a Go To option that enables you to freely surf the Web. You can use this Go To option to test Web pages for TV compatibility as you are developing them. It is good practice when creating Web pages on the computer to periodically post those pages to the external Web server and view them through the WebTV Plus service.

To view the Web pages you create over the WebTV Plus service:

1. Power on the WebTV Plus Receiver.

2. Once the WebTV Plus service is started, press the View button on your WebTV remote control or infrared keyboard.

3. Press Options on the WebTV remote control or infrared keyboard.

4. Press the Go To button, and then type in the URL of the Web page you want to view.

5. Press the Return key on the keyboard or the Go key on the remote control.

   NOTE The WebTV Plus Receiver caches Web pages when they are downloaded to the set-top box. Whenever you post an improved version of a page, be sure to use the Option Dialog Reload button on every page involved to make sure you are seeing your most current work.

SETUP FOR THE WEBTV VIEWER

The WebTV Viewer 2.0 is a software tool that simulates the WebTV Plus service. The WebTV Viewer:

- Shows what happens when pages are compressed horizontally to fit the WebTV screen width of 544 pixels.

- Shows how pages look after the WebTV Viewer enlarges HTML text to the user's text settings, usually a large-sized Helvetica font to increase readability on the TV screen.

- Supports the same HTML tags, and most of the same multimedia technologies, as WebTV.

The WebTV Viewer does not:

- Reproduce TV screen problems such as hard-to-read or small text.

- Reproduce distortion caused by bright colors such as red and white.

- Support any sound capabilities.

Because of its limitations, the WebTV Viewer is a good tool for interim testing of Web pages designed for TV. It should not, however, replace the practice of posting pages to a Web server and viewing them through the actual WebTV service on a WebTV Network or Microsoft TV set-top box.

To install the WebTV Viewer:

1. Open Microsoft Windows Explorer on your computer by clicking Start. Point to Programs, and click Windows Explorer.

2. In Windows Explorer, create a new folder under the C: drive and name it WebTV Viewer.

3. Connect to the Internet, open Internet Explorer and navigate to *http://developer.webtv.net/design/tools/viewer/*.

4. Download the WebTV Viewer Setup.exe file to the folder you created in Step 2.

5. Run the Setup.exe program, and follow on-screen prompts to install the WebTV Viewer.

6. To run the WebTV simulator, click Start, point to Programs, and click WebTV Viewer.

7. The WebTV Viewer displays Web pages on the personal computer in much the same way as they will appear when shown on TV over the WebTV system, as shown in Figure 3.2.

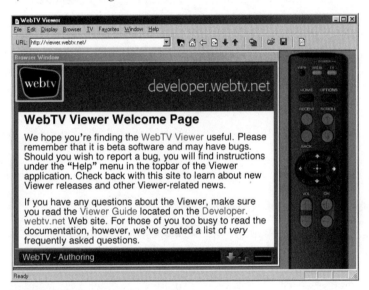

Figure 3.2 *The WebTV Viewer.*

The WebTV Viewer behaves much like the WebTV Plus Receiver. Even the remote control works! For example, if you are connected to the Internet, you may navigate to a Web page like *http://www.microsoft.com/tv/itvsamples*, where sample applications for this book are hosted.

To navigate with the WebTV Viewer:

1. Connect to the Internet if you are not connected already.

2. In the WebTV Viewer, click Options on the remote control. To view the remote control, select Remote Control on the View menu.

3. Click Go To.

4. Type any valid URL into the Address box, and then click Go To Page or press the Enter key on your computer keyboard.

   NOTE   You may view files accessible from the hard disk drive or on a CD by clicking Open from the File menu and then selecting the Browse button.

SETUP FOR PWS

If you are part of a LAN but have direct access to only a single Windows 98 machine, you may wish to use PWS to simulate a Windows NT IIS Web server on your desk. With this setup, you can create and test ASP and connect to back-end data stores using Data Source Name (DSN) and ActiveX Data Objects (ADO). If you use Microsoft Windows 98, you need to install PWS to use ASP or ADO technology and get this Web server capability.

To install PWS:

1. Insert your Windows 98 CD-ROM in the drive on your computer.

2. Click Start, and then click Run to open the Run dialog box.

3. Type D:\add-ons\pws\setup.exe, where D is the letter representing your CD-ROM drive.

4. Click OK to begin the setup process. Follow the directions in PWS setup.

Testing PWS on Windows 98

The installation of PWS on your development machine automatically creates a new directory named Inetpub on your C: drive. Create a new folder under the Wwwroot folder in Inetpub and name it HumIns. From the companion CD, select Microsoft TV E-Commerce, select Humongous Insurance, and then follow the instructions on the Humongous Insurance page to copy the HumIns folder to the folder in Wwwroot that you just created. The C:\Inetpub\Wwwroot directory structure is used on PWS–based and IIS-based Web servers.

   NOTE   The files in the HumIns folder are read-only. If you want to modify the files at a later time, you must uncheck the Read-only attribute for the files. To uncheck the Read-only attribute for the files in the HumIns folder, double-click the HumIns folder in Wwwroot. From the Edit menu in Windows Explorer, click Select All. Right-click on a file in the Windows Explorer, and then click Properties from the shortcut menu. Uncheck the Read-only option, and then click OK.

To test PWS, you need to know your computer's name for a multicomputer setup or the Internet Protocol (IP) address number for a single computer setup.

To determine the name of your development computer:

1. Open the Network dialog box by right-clicking Network Neighborhood on your Windows 98 desktop, and then click Properties.

2. Click the Identification tab in the Network dialog box, and make a note of the value in the Computer Name text box. That is your computer's name.

To determine your computer's IP address:

1. To open an MS-DOS window, click Start, point at Programs, and click MS-DOS Prompt.

2. Type ipconfig at the C:\Windows> prompt to get configuration details for your system.

3. Make note of the top set of four numbers and periods, known as octets, across from IP Address. Your computer's IP address will look something like 172.33.90.145.

To open base.html on the PWS

1. Open your standard Web browser (not the WebTV Viewer) and type http://Mycomputer/HumIns/base.html in the Address text box, where the value of Mycomputer is the IP address or name of your computer. Loading base.html will take a few seconds because your machine is acting as both client and server. The performance of PWS is drastically improved by having separate machines for client and server functionality. The performance of PWS on a machine acting as both client and server may be improved by using the IP address rather than machine name in Microsoft Internet Explorer, as well as the WebTV Viewer.

2. In the Humongous Insurance form on your Web browser, enter a value in the Name text box. You can enter your own name or a made-up name like Joe Interactive. Enter any e-mail address; for example, joeI@itv_jocks.com. Make sure "E-mail me a quote" is selected in the Select Option Group and click Submit to see a dynamically rendered ASP page.

    NOTE To test ASP-ADO pages in the WebTV Viewer, use your computer's IP address rather than its name. For example, if your IP address is 172.30.90.142, you would type http://172.30.90.142/humins/base.html in the URL text box and press Enter to see the Humongous Insurance demo on the WebTV Viewer.

Thanks to PWS, your Microsoft Windows 98 machine is now a Web server! PWS even works across an intranet. If you are connected to a LAN, any other machine on the LAN can view Web pages from your Web server. This enables you to develop e-commerce applications using ASP-ADO code and test them before posting them to your ISP Web server. For more information on creating content using ASP, see chapters 17–20 of this book.

THE ULTIMATE INTERACTIVE TV STUDIO

If your budget allows it, you may want to create the ultimate interactive TV studio, as shown in Figure 3.3.

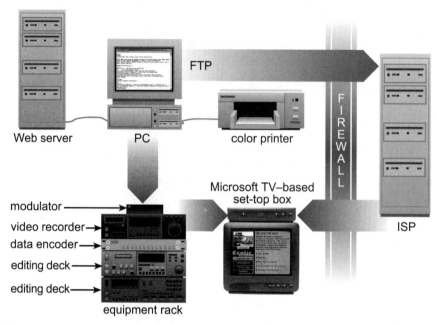

Figure 3.3 *The ultimate interactive TV studio.*

Unlike the recommended setup covered earlier in this chapter, the ultimate interactive TV studio contains an equipment rack that holds a data encoder, a modulator, videotape machines of varying type and quality, and a switching system to mix and match inputs with outputs.

> **NOTE**   If your organization does not already own a data encoder, this section, most likely, will not apply to you. For example, if you are part of a Web development team charged with creating the HTML portion of an interactive TV show, setting up a vertical blanking interval (VBI) encoder is probably not in your best interest. A television station or production studio, or a third-party vendor that specializes in inserting data into video, would probably handle the task. In fact, even television studios generally use a third-party vendor to insert data into the video signal.

The following are the possible components of the equipment rack:

■ **Data Encoder**—The TES3 and TES5 encoders from Norpak Corporation represent the vast majority of commercial quality video encoders in use today. Current prices for Norpak encoders start at about $3,500.

  ❑ **TES3**—For analog networks (most cases), order this model.

- ❑ **TES5**—For digital head ends, order this model.

- ❑ **EIA-516 NABTS Data Broadcast Software**—This module enables a TES3 or TES5 to insert interactive content into a video signal on lines 10–20 of the VBI.

- ❑ **EIA-608 Caption Encoder Software**—This module enables a TES3 or TES5 encoder to provide closed-captioning support. In more technical terms, it enables a TES3 or TES5 encoder to insert data into line 21 of the VBI of a video signal.

- ■ **Tape Decks**—The tape decks will be some combination of VHS, High 8, and studio-quality Beta. Video recorders are manufactured by a wide variety of companies. Prices range from a few hundred dollars for home versions to tens of thousands of dollars for a studio-quality machine.

- ■ **Modulator**—The modulator tunes the video signal to a specific channel for viewing on TV.

For more information about third-party vendors who can encode HTML content onto video for you, see the "Interactive TV Vendors" section in Chapter 15, "Fundamentals of Delivering Interactive TV Content."

How the Components of the Equipment Rack Fit Together

Figure 3.4 shows a simplified diagram of how the components of the equipment rack—including the data encoder, modulator, and video recorder—work together for encoding and recording interactive TV content.

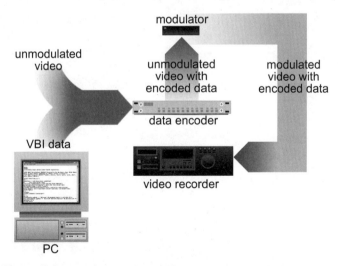

Figure 3.4 *Encoding and recording interactive TV content.*

As shown in Figure 3.4, a Windows 98, Windows NT, or Windows 2000 computer containing data that is to be encoded into a video signal is connected to the data port on the encoder. The connection between PC and video encoder may be though a serial cable connecting COM ports or through an Ethernet LAN connection, depending on the make and model of the VBI encoder.

A video feed is connected to the video In port on the VBI encoder. For test purposes, the video feed may come right from a cable service provider. In practice, the feed will normally come from the video Out port on one of the recorders in the equipment rack.

The video Out port of the VBI encoder is connected to the video In port on the modulator. This connection carries video with data from the computer encoded into the lines of the VBI. Encoded data may be as simple as a link to interactive TV content carried on text 2, line 21 of the VBI or as complex as TCP/IP packets containing all the HTML coding for the interactive TV experience. For more information about the VBI, see Chapter 15, "Fundamentals of Delivering Interactive TV Content."

The modulated signal may be set to a particular television channel for viewing on a WebTV or Microsoft TV receiver or straight out the broadcast head end. More commonly, the combined video and data signal is recorded back onto tape using one of the video recorders in the equipment rack as a target.

The amount and type of data encoded into the video signal determines what sort of recorder is used as the target. If line 21 interactive TV links only are encoded into the video stream, a standard VHS tape deck is adequate. If TCP/IP packets containing all the data that make up an interactive TV experience are to be included in the video stream, a more expensive High 8 or professional-quality Beta deck is used.

The vast majority of interactive TV programming is delivered using interactive TV links on line 21 of the VBI. The only time TCP/IP packet data is delivered over the VBI is when it is delivered to the small number of Advanced Television Enhancement Forum (ATVEF) receivers that have no computer or cable modem back channel.

WHAT'S NEXT

This chapter provided an overview of the hardware and software needed for creating interactive TV content. For updated information about hardware and software, as well as a list of vendors involved in interactive TV, see *http://www.microsoft.com/tv*. In the next chapter, "Fast Track for Creating Microsoft TV Content," we will discuss how to use the equipment described in this chapter to create content for Microsoft TV.

Chapter 4

Fast Track for Creating Microsoft TV Content

In This Chapter

- Where to Find Sample Content for This Chapter

- Copying the Source Files

- Overview of the Template_main.html Page

- Modifying the Template_tv.html Page

- The TV Object and Full Screen Button

- Modifying the Template_option1.html Page

- Testing Content on the WebTV Viewer

- Testing the Page on TV

Designing interactive TV content is really not much different from designing Web pages, although there are a few tricks you need to know about to handle the display differences between the PC and the TV. To help you through the rough spots and to make sure you get your own interactive TV content up and running quickly, we've provided a generic template that you can customize for almost any type of interactive TV content. This chapter shows how you can transform the generic template into an "information-on-demand" application for a news broadcast, enabling TV viewers to access news headlines, weather, sports scores, or stock quotes with a click of their remote.

Using the generic template as a starting point for creating interactive content provides several important advantages. For starters, it follows Advanced Television Enhancement Forum (ATVEF) content creation standards, so you can be assured that the template will run across all interactive TV receivers that follow ATVEF standards. Another big advantage of the generic template is that it dynamically applies separate style sheets for you to make adjustments between the PC and TV display characteristics. That way, what you see on your PC is almost identical to what you see on the TV. Perhaps most importantly, the generic template saves you time because its page size, colors, and fonts are already optimized for TV.

As a full-service introduction to designing interactive TV content, this chapter also describes how to test content on the WebTV Viewer and how to post the content using FTP software so the content can be tested on TV. When you're done with this chapter, and you've faithfully followed the steps in the tutorial, you'll have a good understanding of the overall interactive TV design process and working interactive TV content that you can demonstrate to others.

WHERE TO FIND SAMPLE CONTENT FOR THIS CHAPTER

This book is designed to be used in conjunction with the *Building Interactive Entertainment and E-Commerce Content for Microsoft TV* companion CD and Web site. In addition to reading this chapter, we recommend that you view the sample files on both of those references.

■ To view the generic template, either on a set-top box running Microsoft TV or on the Web, go to the *http://www.microsoft.com/tv/itvsamples*.

■ To view the sample files associated with this chapter, refer to the "Fast Track for Creating Interactive TV Content" topic in the "Microsoft TV Primer" section of the companion CD.

■ For HTML source files in the generic template, see the "Templates" section of the companion CD.

COPYING THE SOURCE FILES

To begin modifying the template, copy the generic folder on the companion CD, which contains the HTML source files and graphics for the generic template, to your computer's hard drive.

1. To copy the generic folder, start the companion CD by placing it into your CD-ROM drive and then clicking *Building Interactive Entertainment and E-Commerce Content for Microsoft TV* CD-ROM. Navigate to the file Default.html and then double-click the file.

2. In the left navigation panel, click Templates.

3. Open Microsoft Windows Explorer.

4. In Microsoft Windows 98 and Windows 2000, double-click the My Computer icon on the Desktop.

5. For Windows 95 and Windows NT, click Start on the taskbar, click Programs, and then double-click Explorer.

6. Activate the companion CD. Using the right mouse button, click and drag the generic folder from the Templates box to your hard drive, and then select Copy Here from the shortcut menu.

Uncheck the Read-Only Attribute for the Files

The files in the Templates folder are read-only. To modify the files, you must uncheck their read-only attribute. To uncheck the read-only attribute for the files, double-click the Templates folder you just copied. From the Edit menu in Windows Explorer, click Select All. Right-click on a file in Windows Explorer, and then click Properties from the shortcut menu. Uncheck the read-only option, and then click OK.

Overview of the Generic Interactive TV Template

To view the generic template in Microsoft Internet Explorer, on your hard drive, double-click the file template_main.html in the generic folder to open the template in Internet Explorer, as shown in Figure 4.1.

The generic template is a frame-based page that consists of the following types of pages:

- **template_main.html**—defines the basic structure for the template.

- **template_tv.html**—hosts the TV object and the main navigation controls.

- **template_option*xx*.html**—content pages that are dynamically swapped out of the content frame when the user selects links, which may be either on the template_tv.html page or one of the options pages. An options page, customized for news content, is shown in Figure 4.2.

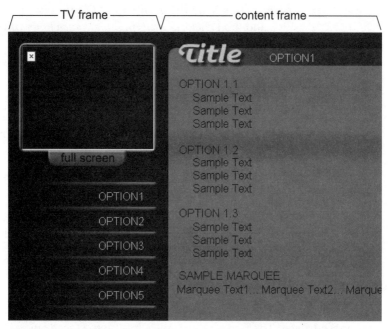

Figure 4.1 *The generic template.*

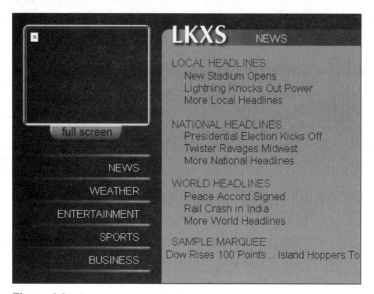

Figure 4.2 *The generic template customized for news content.*

OVERVIEW OF THE TEMPLATE_MAIN.HTML PAGE

The template_main.html page is the basic frameset page that determines the structure of the generic template. In our example, template_main.html page consists of a frameset that is divided into two frames: the TV frame and the content frame. This two-frame structure serves as the foundation for almost all of the interactive TV content in this book. For this tutorial, the template_main.html page does not require modification, but we will review the source code to get an idea how it works.

To view the HTML code for the template_main.html Page:

To better understand how the template works, view the HTML code in the Internet Explorer window. Open the template_main.html page in Internet Explorer, click the View menu, and then click Source.

```
<!-- JavaScript removes the grey borders from around a frame for WebTV. -->
<script type="text/JavaScript">
if (navigator.appVersion.indexOf("WebTV") != -1 )
{
    document.write("<body hspace=0 vspace=0>");
}
</script>
<!-- frames -->
<frameset cols="242,*" framespacing=0 frameborder=0
hspace=0 vspace=0 marginwidth=0 marginheight=0>
<frame name="tv" src="template_tv.html" marginwidth="0"
marginheight="0" scrolling="no" noresize frameborder="no">
<frame name="content" src="template_option1.html"
marginwidth="0" marginheight="0" scrolling="no" noresize frameborder="no">
</frameset>
```

Now close the Notepad window that contains the source code window for template_main.html.

MODIFYING THE TEMPLATE_TV.HTML PAGE

To open the template_tv.html page in Internet Explorer, click Open from the File menu of the Internet Explorer window. Click Browse, and then select the template_tv.html file from the generic folder on your hard drive. The template_tv.html page, shown in Figure 4.3, is the left TV frame of the template_main.html page and contains the TV object that displays the TV picture and the navigation controls that dynamically swap out pages in the content frame.

To view the source code for the template_tv.html file, click the View menu in Internet Explorer, and then click Source.

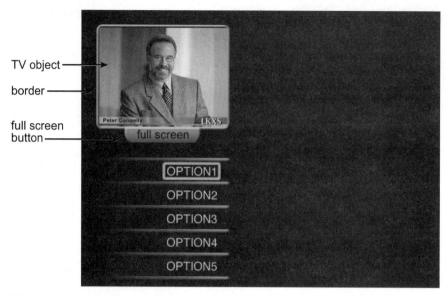

Figure 4.3 *The template_tv.html page.*

The Style Sheet Section of Code

The template_tv.html page has been formatted with two style sheets: one for the computer and one for the TV. The following code sample illustrates that JavaScript is used to detect whether Internet Explorer is located on the computer. If Internet Explorer is detected, the cssPC.css stylesheet automatically adjusts the font sizes so that what you see on the computer is nearly identical to what you see on TV. Television requires larger fonts, such as 16 point, so that viewers can read the text from across the room. However, if you specify 16-point text on a computer page, the text just fits on the page. The dual style sheet solution, as shown in the following sample code, solves this problem by adjusting the font sizes for the appropriate device.

```
<html>
<head>
<link rel="stylesheet" type="text/css" href="cssTV.css">
<script language="JavaScript">
var bName = navigator.appName
if (bName == "Microsoft Internet Explorer")
{
    document.write("<link rel='stylesheet' type='text/
css' href='cssPC.css'>");
}
</script>
```

THE TV OBJECT AND FULL SCREEN BUTTON

Probably the most important element on the template_tv.html page is the TV object. The TV object is used to display the TV picture while the user interacts with the Web content. In this way, the user never loses the continuity with the television programming. As shown in the following sample code, the TV object is positioned on top of a border for aesthetics, using an <img> tag. The TV object is created by specifying *data="tv:"* in the <object> tag. The Full Screen button, in turn, is implemented as an <img> tag, with the <a href="tv;"> anchor defined so that when the image is clicked, full-screen TV is restored.

```
<body bgcolor="#1E1032" topmargin=0 leftmargin=0">
<div style="position:absolute;top:16;left:16">
<img src="images/border3.gif" width=210 height=160 alt="" border="0">
</div>
<div style="position:absolute;top:21;left:21">
<object data="tv:" width=200 height=150 alt="" border="0">
</object>
</div>
<div style="position:absolute;top:175;left:61">
<a href="tv:"><img src="images/full3.gif" width=119 height=23 alt=""
border="0"></a>
</div>
```

Navigation for the Page

The navigation for the template_tv.html page is handled through links that are defined for text, as shown in bold in the following code sample. Notice that the target for the links is the content frame and that the code is a combination of tables and divisions. Absolute positioning gives you great fidelity on positioning elements, and enclosing text and graphics in a table makes it easy to move the content as a block.

```
<div style="position:absolute;top:218;left:16">
<table width=210 cellspacing=0 cellpadding=0 border=0>
<tr>
<td height=8 valign=top colspan=2><img src="images/line.gif"
width=210 height=8 alt="" border="0" hspace="0" vspace="0"></td>
</tr>
<tr>
<td valign="middle" align="right" height=28 width=190>
<a href="template_option1.html" target="content">
<font class="clsButtontext">OPTION1</font></a></td>
<td width=20> </td>
</tr>
```

Changing the Link Text

Now that we have reviewed the structure of the page, let's modify the page, beginning with the link text. In the HTML code, change the text of the links to the text shown in bold in the following sample code. For example, change OPTION1 to NEWS, OPTION2 to WEATHER, and so on.

```
<div style="position:absolute;top:218;left:16">
<table width=210 cellspacing=0 cellpadding=0 border=0>
<tr>
<td height=8 valign=top colspan=2><img src="images/line.gif"
width=210 height=8 alt="" border="0" hspace="0" vspace="0"></td>
</tr>
<tr>
<td valign="middle" align="right" height=28 width=190>
<a href="template_option1.html" target="content">
<font class="clsButtontext">NEWS</font></a></td>
<td width=20> </td>
</tr>
<tr>
<td height=8 valign=top colspan=2><img src="images/line.gif" width=210
height=8 alt="" border="0" hspace="0" vspace="0"></td>
</tr>
<tr>
<td valign="middle" align="right" height=28 width=190>
<a href="template_option2.html" target="content">
<font class="clsButtontext">WEATHER</font></a></td>
<td width=20> </td>
</tr>
<tr>
<td height=8 valign=top colspan=2><img src="images/line.gif"
width=210 height=8 alt="" border="0" hspace="0" vspace="0"></td>
</tr>
<tr>
<td valign="middle" align="right" height=28 width=190>
<a href="template_option3.html" target="content">
<font class="clsButtontext">ENTERTAINMENT</font></a></td>
<td width=20> </td>
</tr>
<tr>
<td height=8 valign=top colspan=2><img src="images/line.gif"
width=210 height=8 alt="" border="0" hspace="0" vspace="0"></td>
</tr>
<tr>
<td valign="middle=" align="right" height=28 width=190>
<a href="template_option4.html" target="content">
<font class="clsButtontext">SPORTS</font></a></td>
<td width=20> </td>
```

```
</tr>
<tr>
<td height=8 valign=top colspan=2><img src="images/line.gif"
width=210 height=8 alt="" border="0=" hspace="0" vspace="0"></td>
</tr>
<tr>
<td valign="middle" align="right" height=28 width=190>
<a href="template_option5.html" target="content">
<font class="clsButtontext">BUSINESS</font></a></td>
<td width=20> </td>
</tr>
<tr>
<td height=8 valign=top colspan=2><img src="images/line.gif"
width=210 height=8 alt="" border="0" hspace="0" vspace="0"></td>
</tr>
</table>
</div>
```

Saving the Page

Once you have changed the text, click Save from the File menu. Close the source code window, switch to the Internet Explorer window, and click Refresh from the View menu. The file should now look like Figure 4.4.

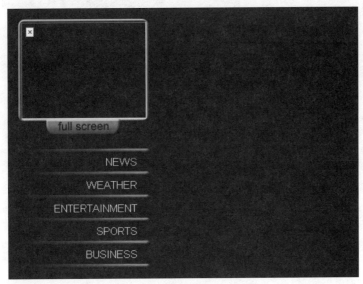

Figure 4.4 *The modified template_tv.html page.*

Now close the Notepad window that contains the source code window for template_tv.html.

MODIFYING THE TEMPLATE_OPTION1.HTML PAGE

To open the template_option1.html page in Internet Explorer, click Open from the File menu. Click Browse, and then select the template_option1.html file from the generic folder on your hard drive. The template_option1.html page, shown in Figure 4.5, is the default page for the content frame. This page consists of a logo, several suboption areas, and a scrolling marquee at the bottom of the page.

However, before modifying any code on the template_option1.html page, review various portions of the code in order to see their purpose. To view the HTML source code for the file template_option1.html, click Source on the View menu of Internet Explorer.

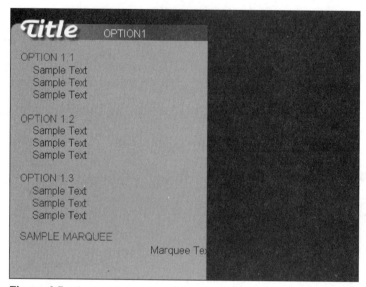

Figure 4.5 *The template_option1.html page.*

The Body Background, Content Background, and Logo

As you can see by the following sample code, the background color for the page is set by specifying the bgcolor="#0c073b" attribute for the <body> tag. The content area color is set by specifying "background-color:#9992B2" for the style property of a <div> tag. The logo, specified with the <img> tag, is then layered over the division.

```
<!--sets the background color and margins for the page-->
<body bgcolor="#0c073b" topmargin=0 leftmargin=0>
<!--sets the content background for the page-->
<div style="position:absolute;top:16;left:0;
background-color:#9992B2;width:318;height:404">
```

```
<!--specifies the logo image-->
<div style="position:absolute;top:0;left:0">
<img src="images/title.gif" width=302 height=34 alt="" border="0">
</div>
```

The Options Sections

The Options sections, as shown in the sample code, are designed for easy modification so you can quickly change the text, create anchors for the options, or add and remove options as needed.

```
<!--Heading for the first table-->
<div style="position:absolute;top:50;left:16">
<table cellpadding=0 cellspacing=0 border=0 width=302>
<tr>
<td  align=left valign=middle><font class="clsHeadtext">
OPTION 1.1</font></td>
</tr>
</table>
</div>

<!--Suboptions for the first table-->
<div style="position:absolute;top:70;left:36">
<table cellpadding=0 cellspacing=0 border=0 width=302>
<tr>
<td align=left valign=middle><a href="template_option1a.html"
target="content">Sample Text</a></td>
</tr>
```

The Marquee

The marquee section of the document, implemented with the following code, enables you to define scrolling text (a ticker) for the page:

```
<!--Heading for the Marquee table-->
<div style="position:absolute;top:329;left:16">
<table cellpadding=0 cellspacing=0 border=0 width=302>
<tr>
<td  align=left valign=middle><font class="clsHeadtext">SAMPLE
MARQUEE</font></td>
</tr>
</table>
</div>

<!--Text for Marquee-->
<div style="position:absolute;top:349;left:0">
<marquee direction="left" height=20 width=318 scrolldelay=100>
<tr>
```

(continued)

```
<td><a href="marquee.htm"><font class="clsDescription">Marquee Text1...
</font></a></td>
</tr>
<tr>
<td><a href="marquee.htm"><font class="clsDescription">Marquee Text2...
</font></a></td>
</tr>
<tr>
<td><a href="marquee.htm"><font class="clsDescription">Marquee Text3...
</font></a></td>
</tr>
</marquee>
</div>
```

Changing the Logo

Change the logo for the template_option1.html page by specifying a different image. To do this, change src="images/title.gif" to src="images/lkxs.gif". The code should now look like the following sample, as shown in bold:

```
<!--specifies the logo image-->
<div style="position:absolute;top:16;left:0">
<img src="images/lkxs.gif" width=302 height=34 alt="" border="0">
</div>
```

Changing the Options Headings

Create the news headlines in the template_tv.html page by changing the link text to the bold text in the following code sample.

```
<div style="position:absolute;top:66;left:16">
<table cellpadding=0 cellspacing=0 border=0 width=302>
<tr>
<td  align=left valign=middle><font class="clsHeadtext">LOCAL
HEADLINES</font></td>
</tr>
</table>
</div>

<div style="position:absolute;top:86;left:36">
<table cellpadding=0 cellspacing=0 border=0 width=302>
<tr>
<td align=left valign=middle><font class="clsDescription">New Stadium
Opens</font></a></td>
</tr>
<tr>
<td align=left valign=middle><font class="clsDescription">Lightning
Knocks Out Power</font></td>
</tr>
```

```
<td align=left valign=middle><font class="clsDescription">More Local
Headlines...</font></td>
</tr>
</table>
</div>

<div style="position:absolute;top:160;left:16">
<table cellpadding=0 cellspacing=0 border=0 width=302>
<tr>
<td  align=left valign=middle><font class="clsHeadtext">NATIONAL
HEADLINES</font></td>
</tr>
</table>
</div>

<div style="position:absolute;top:180;left:36">
<table cellpadding=0 cellspacing=0 border=0 width=302>
<tr>
<td align=left valign=middle><font
class="clsDescription">Presidential Election Kicks Off</font></td>
</tr>
<tr>
<td align=left valign=middle><font class="clsDescription">Twisters
Ravage Midwest</font></td>
</tr>
<tr>
<td align=left valign=middle><font class="clsDescription">More
National Headlines...</font></td>
</tr>
</table>
</div>

<!--Heading for third table-->
<div style="position:absolute;top:238;left:16">
<table cellpadding=0 cellspacing=0 border=0 width=302>
<tr>
<td  align=left valign=middle><font class="clsHeadtext">WORLD
HEADLINES</font></td>
</tr>
</table>
</div>

<!--Suboptions for the third table-->
<div style="position:absolute;top:258;left:36">
<table cellpadding=0 cellspacing=0 border=0 width=302>
<tr>
```

(continued)

```
<td align=left valign=middle<font class="clsDescription">Peace Accord
Signed</font></a></td>
</tr>
<tr>
<td align=left valign=middle><font class="clsDescription">Rail Crash
in India</font></td>
</tr>
<tr>
<td align=left valign=middle><font class="clsDescription">More World
Headlines...</font></td>
</tr>
</table>
</div>
```

Changing the Marquee Text

Now change the marquee text to the bold text in the following sample code:

```
<div style="position:absolute;top:345;left:16">
<table cellpadding=0 cellspacing=0 border=0 width=302>
<tr>
<td align=left valign=middle><font class="clsHeadtext">BREAKING
NEWS</font></td>
</tr>
</table>
</div>

<div style="position:absolute;top:365;left:0">
<marquee direction="left" height=20 width=318 scrolldelay=100>
<tr>
<td><a href="marquee.htm"><font class="clsDescription">
Antitrust Trial Closes to an End...</font></a></td>
</tr>
<tr>
<td><a href="marquee.htm"><font class="clsDescription">
Island Hoppers Top MudHens...
</font></a></td></tr>
<tr>
<td><a href="marquee.htm"><font class="clsDescription">New Stadium Opens...
</font></a></td>
</tr>
</marquee>
</div>
```

Saving the Page

After you have changed the text, click Save from the File menu. Close the source code window, activate the Internet Explorer window, and click Refresh from the View menu. The file should now look like Figure 4.6.

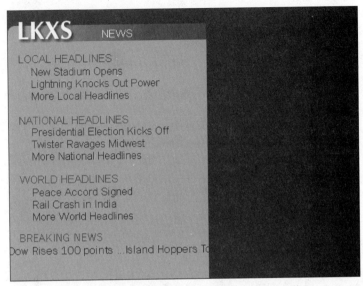

Figure 4.6 *The template_option1.html page.*

Now close the Notepad window that contains the source code window for template_option1.html. You can leave the Internet Explorer window open if you like.

TESTING CONTENT ON THE WEBTV VIEWER

The WebTV Viewer simulates a Microsoft TV–based set-top box, rendering content in much the same way as you would see it on TV. If you have not already installed the Viewer, do so now by following the instructions in the "Setup for the WebTV Viewer" section of Chapter 3.

About Web Mode and TV Mode

There are two modes used by Microsoft TV:

■ **Web mode**—the mode used by Microsoft TV for standard Web pages. In Web mode, traditional Web functions such as Go To, Save, Send, and Print are available. Those Web pages that do not contain the video object are shown in Web mode.

■ **TV mode**—a special mode used for showing interactive TV content that contains the video object. When pages are shown in TV mode, the TV picture appears in the video object, the page cannot be scrolled, and TV functions such as Today's Listings, Program Info, and TV Favorites are available in place of Web functions.

For more detail about the two modes, refer to Chapter 5, "Guidelines for Designing Microsoft TV Content."

The viewset.html Page

Each *Building Interactive Entertainment and E-Commerce Content for Microsoft TV* template folder contains a viewset.html file that simulates the interactive TV link that appears during a television program or commercial. The viewset.html page, shown in Figure 4.7, features a Go Interactive button that, when clicked, forces the WebTV Viewer into TV mode, so that the TV picture appears in the TV object. (The WebTV Viewer simulates the TV object by placing a graphic image in the TV object area.)

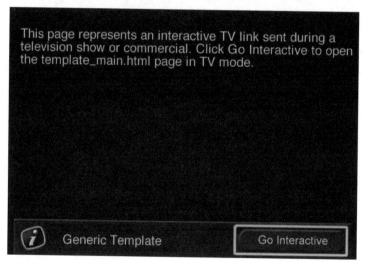

Figure 4.7 *The viewset.html page as shown on TV.*

To view the modified content in TV mode, open the file viewset.html (located in the generic folder on your hard drive) in the WebTV Viewer.

1. Start the WebTV Viewer, and then click Open File from the File menu.

2. Double-click viewset.html, located in the generic folder on your hard drive.

3. Click the Go Interactive button in the WebTV Viewer window.

   The window should show an image in the TV object space, as shown in Figure 4.8.

Figure 4.8 *The template_main.html page as shown in the WebTV Viewer.*

TESTING THE PAGE ON TV

To test your modified content on an interactive TV set-top box, you must first post the files to an external Web server and then simulate an interactive TV link by using the viewset.html file located in the generic folder.

Posting the Files

To perform a final test, you must post the tutorial files, including the image folder, to an external Web server using FTP software. For this book, a WebTV-based Internet terminal/receiver is used as the primary target browser; therefore, the following instructions apply specifically to the WebTV-based Internet terminal/receiver. To view the page on Microsoft TV:

1. Start the WebTV service, click Options, and then click Go To.

2. In the address box, type the URL of the viewset.html file you just posted.

3. On the Microsoft TV-based Internet terminal/receiver remote control, click the Go button or press Enter on the keyboard to select the Go Interactive button.

4. You should now see your interactive content with the TV picture in the TV object space. To return to full-screen TV, select the Full Screen button underneath the video object. To return to the viewset page, click the View button on the remote control.

WHAT'S NEXT

This chapter was designed to get you started designing interactive TV content. In the following chapters, you will find a series of easy-to-follow tutorials and proven strategies for designing interactive TV content for Microsoft TV.

Part II

Microsoft TV Design Guide

```
var oldsrc = ""

function featureload()
{
        parent.frames["content"].location.href="exotic_feature1.html";
        document.btn_Feature.src = featuredon.src;
        if (oldimage != document.btn_feature)
        {
                oldimage.src = oldsrc
        }
        oldimage = document.btn_Feature
        oldsrc = featuredoff.src
}

function reservationsload()
{
        parent.frames["content"].location.href="exotic_reservations1.html";
        document.btn_reservations.src = reservationson.src
        if (oldimage != document.btn_reservations)
        {
                oldimage.src = oldsrc
        }
        oldimage = document.btn_reservations
        oldsrc = reservationsoff.src
```

Chapter 5

Guidelines for Designing Microsoft TV Content

In This Chapter

■ Using the Companion CD and Web Site

■ Fitting Content into Microsoft TV's Design Area

■ Web Mode vs. TV Mode

■ Strategies for Designing Content

Designing interactive TV content requires a certain amount of discipline, but you do not need to be an HTML guru to do it. By following the instructions in the "Microsoft TV Design Guide" part of this book, anyone with an intermediate-level understanding of HTML can create interactive TV content. The strategies, methods, and techniques presented here will get you quickly up to speed on creating interactive TV content.

This chapter lays the foundation for interactive TV design. The strategies shown here will save you time and effort and help ensure that the content you create works not only on Microsoft TV, but also on other TV receivers built to Advanced Television Enhancement Forum (ATVEF) standards.

USING THE COMPANION CD AND WEB SITE

The *Building Interactive Entertainment and E-Commerce Content for Microsoft TV* companion CD and Web site work hand-in-glove with the text in this book to provide designers with hands-on experience. The following sections of the companion CD will be of particular value to designers:

- **Microsoft TV Design Guide**—This guide contains a wide variety of coding and user interface ideas. You can use its content as a starting point for building your own interactive TV content. At the beginning of nearly every chapter in the guide, you will find a "Where to Find Sample Content" section, which gives the location of sample pages and source code associated with the chapter. The samples are also posted for viewing over Microsoft TV at *http://www.microsoft.com/tv/itvsamples.*

- **Microsoft TV Programmer's Guide**—This section of the companion CD provides comprehensive coverage of the Microsoft TV object model, including the objects, properties, methods, and events supported by Microsoft TV. It also covers the HTML, Dynamic HTML (DHTML), and Cascading Style Sheets level 1 (CSS1) support provided by Microsoft TV. If you have questions about which objects, elements, CSS1 properties, or scripting languages are supported by Microsoft TV, you will find the answers here.

- **Templates**—This portion of the companion CD consists of a series of interactive TV folders. You can drag these folders from the companion CD to your hard drive and use them as a starting point for your own interactive TV content.

Another valuable resource is this book's companion Web site, which contains sample content that you can view over Microsoft TV and other TV receivers built to ATVEF standards. You can find the companion Web site at *http://www.microsoft.com/ tv/itvsamples.*

FITTING CONTENT INTO MICROSOFT TV'S DESIGN AREA

Perhaps the biggest challenge in designing HTML content for TV is fitting all the information you want within the confines of Microsoft TV's 560-pixel-wide by 420-pixel-high screen dimensions. Web pages in TV mode on Microsoft TV are not scrollable, so it is essential that interactive TV content fits within Microsoft TV's design area. Figure 5.1 shows Microsoft TV's screen dimensions.

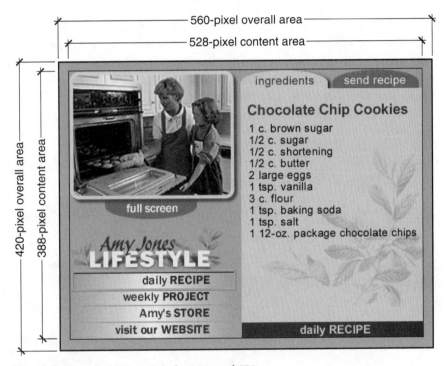

Figure 5.1 *Screen dimensions for Microsoft TV.*

The above illustration reflects a 16-pixel designer margin. By default, Microsoft TV imposes margins on content to ensure that it will look good on TV. For example, Microsoft TV automatically creates a top margin of 6 pixels and a left margin of 8 pixels. We recommend a 16-pixel margin for content, both because this is aesthetically pleasing and because it ensures that older TV sets with rounded corners will not cut off critical content. Accordingly, all interactive TV content in this book and on the companion CD is designed to fit within a 528-pixel-wide by 388-pixel-high design space.

WEB MODE VS. TV MODE

Microsoft TV–based set-top boxes have two modes: Web mode and TV mode. Both modes have the same screen dimensions, but they offer different functionality. When developing interactive TV content, you must design content for TV mode. However, it is a good idea to be aware of the capabilities and limitations of both modes, as described in the following sections.

Web Mode

Web mode is the mode used by Microsoft TV viewers to surf the Web and perform traditional Internet-related functions such as page scrolling, printing, and saving documents. In Web mode, if users click the Options button on their remote control, they see an Options panel that they can use to surf the Web, find documents, save documents, or print documents. They can also click the TV Window button on the screen to display a small TV image of the current channel on top of the document. An example of Web mode is shown in Figure 5.2.

Figure 5.2 *Microsoft TV in Web mode.*

TV Mode

TV mode is the mode used by Microsoft TV viewers to watch TV. Therefore, interactive TV content developers must develop their content for TV mode. In TV mode, the TV picture is visible, but content on the screen is not scrollable. When viewers press the Options button on their remote control, they are presented with a series of TV-related options, as shown in Figure 5.3.

No Information   5 LKXS

Today's Listings   Program Info   TV Favorites

Figure 5.3 *Microsoft TV in TV mode.*

Note that the above illustration shows TV mode without any interactive content. For more information on setting TV mode at design time, see Chapter 16, "Creating Interactive TV Links."

STRATEGIES FOR DESIGNING CONTENT

The following sections in this chapter provide a list of proven strategies to follow when designing Microsoft TV content.

Avoid Microsoft TV Proxy Server Conversion

Microsoft TV passes HTML content through a proxy server that applies a series of rules against the content to ensure it will look good on TV. For example:

- Pages wider than 544 pixels are compressed to fit within the 544-pixel width.

- Images wider than 544 pixels are compressed to a 544-pixel width.

- Font sizes are enlarged for TV display, and the font face is converted to Helvetica.

■ Colors that are not compliant with National Television Standards Committee (NTSC) specifications for HTML elements are mapped by the Microsoft TV proxy server to be NTSC–compliant. Note that pure white is tough on the eyes, but the proxy server does not remap it.

■ Scrollable content within frames is converted to a table format, often resulting in a long vertical table.

The best strategy is to design content so that the proxy server makes no conversions on your content at all. The "Microsoft TV Design Guide" is dedicated to that task. As the saying goes, "the best surprise is no surprise at all." The following chapters show you how to design interactive TV content that will pass through the proxy server unchanged, thus protecting the integrity of your design.

Always Include the TV Object

Interactive TV is all about improving the TV experience. Therefore, the TV object, in one form or another, must be integrated into interactive content to provide a seamless transition between traditional full-screen TV and interactive TV. There are two options for including the TV object in interactive TV content: You can layer Web content over a full-screen TV object, or you can layer the TV object over Web content. Figure 5.4 shows the TV object layered over Web content.

Figure 5.4 *The TV object layered over Web content.*

For more information about layering Web content over full-screen TV or layering the TV object over Web content, refer to chapters 6 and 7.

Make the TV Object as Large as Possible

Focus groups, usability studies, and feedback from the television industry have shown that both viewers and producers want the TV object to be as large as possible on interactive TV content. In some cases, you can meet this need by layering interactive TV content over a full-screen TV image. However, in most cases, the TV object must be layered over Web content. In such instances, it is best to err on the side of caution, by making the TV image as large as possible.

Make the TV Object a 4:3 Ratio

NTSC broadcasts use a 4:3 aspect ratio. To avoid distortion of the TV image, you must honor this ratio when adding a TV object to a Web page. An easy way to maintain the 4:3 aspect ratio is to multiply the width of the TV object by .75 to obtain the proper height. For example, if you want to add a TV object that is 250 pixels wide, multiply 250 by .75 to obtain the height of 187 pixels, as shown in the following equation:

```
Tvobjectwidth x .75 = TVobjectheight
250 x .75 = 187.5
```

The following syntax can be used to create a full-screen TV object:

```
<object data="tv:" style="position:absolute;top:0;left:0;z-index:-1"
height=100% width=100%>
```

The Microsoft TV screen dimensions (560 by 420) follow the 4:3 aspect ratio, so the full-screen TV object appears without distortion. For more information about adding the TV object to a Web page, refer to chapters 6 and 7.

Design Content to Be Visible from Across the Room

When designing interactive TV content, keep in mind that TV viewers will typically view the information from across the room. Research indicates that 18-point text is the optimum font size for TV. Nevertheless, design constraints often dictate smaller point sizes. In general, make font sizes as large as possible, while still fitting within the constraints of your overall design. Also, it is best to avoid images with fine detail or embedded text that cannot be distinguished from a distance.

For more information on optimal TV fonts and font sizes, see Chapter 10, "Creating Text for Microsoft TV." For information on creating images for TV, see Chapter 11, "Adding Images to Microsoft TV Content."

Avoid Scrollable Content

When designing interactive TV content, avoid designs that require scrolling. Here are several good reasons why:

- Microsoft usability research indicates that, in general, TV viewers are not accustomed to scrolling. Therefore, if scrollable areas exist in your interactive TV content, viewers may not know how to use them.

- Microsoft TV converts scrollable content in frames and iframes (inline frames) into tables. This conversion changes the height of the frame or iframe, potentially pushing content off the screen.

- In TV mode, Microsoft TV does not support scrollable pages. For example, if content extends beyond the 560-by-420-pixel boundary of Microsoft TV, the content cannot be viewed or accessed by the TV viewer.

For more information about how to design content for TV that does not scroll, see Chapter 7, "Layering TV over Web Content."

Use Absolute Positioning

Strategies for designing interactive TV content will differ, but because the size of a TV page does not change, you can position elements on the page using absolute positioning. Absolute positioning is not considered part of the ATVEF content creation standards, but it is supported by Microsoft TV. Moreover, it makes page design much easier. The following code snippet shows how you can use absolute positioning to place a TV object on a Web page for Microsoft TV:

```
<div style="position:absolute;top:14;left:14;">
<object data="tv:" height=250 width=187>
</div>
```

When using absolute positioning, it's important to fix font sizes with CSS1 properties to prevent the TV viewer from changing the font sizes. For more information about fixing font sizes with CSS1 properties, see Chapter 10, "Creating Text for Microsoft TV."

For more information about absolute positioning, see Chapter 6, "Layering Web Content over Full-Screen TV," and Chapter 8, "Formatting Microsoft TV Content with Styles and Style Sheets."

NOTE   Absolute positioning is not included in the W3C Recommendation for CSS1. As a result, absolute positioning is one area where the sample content in this book deviates from the ATVEF specification.

Strike a Balance Between Visual Appeal and Speed

One of the greatest challenges of designing interactive TV content is finding a balance between visual appeal and the time required for loading pages. Television is, after all, a visual entertainment medium, and TV viewers are accustomed to high-quality television content. As a result, the standard Web page on TV may appear dull and uninviting to the typical TV viewer. However, the visual appeal of a page must be carefully weighed against the time required to load the page. TV viewers expect interactive TV pages to be visually appealing, but they do not want to wait more than a few seconds for the page to load. In the end, there is no easy solution. But as a general guideline, it is best to keep page sizes under 25 KB.

Design Content for Everybody

When creating interactive TV content, design it so it can be used by just about anyone. Nearly every household in North America owns a TV set, but research shows that many interactive users have no experience with a personal computer or Microsoft Windows. As a result, you should design content with simple user interfaces that can be navigated by the technically naive as well as the technically savvy user.

Design Colors to Be NTSC-Safe

With the exception of the color white, the Microsoft TV proxy server automatically maps the colors of HTML elements to NTSC-safe colors (those approved as suitable for TV display by the NTSC), but it is best not to rely on the proxy server to do the work for you. By picking your own NTSC-safe colors, you can minimize any changes imposed by the Microsoft TV proxy server. For more information about selecting colors for Microsoft TV content, see Chapter 9, "Selecting Colors for Microsoft TV Content."

Design Content to ATVEF Guidelines

The foundation of the sample interactive TV pages covered in the Microsoft TV Design Guide is the Specification for Interactive Television produced by the ATVEF. This specification defines the following content creation standards for interactive TV pages: HTML 4.0, CSS1, Level 0 Document Object Model (Level 0 DOM), and the ECMA Language specification (ECMAScript).

The specification for interactive television produced by the ATVEF serves as a catalyst for the development and acceptance of interactive TV technology. For the first time, broadcasters, content developers, and television professionals have a common set of rules—agreed upon by industry leaders—for creating and delivering interactive TV content. Why is it important to follow this specification when designing interactive TV content? Quite simply, when you design content according to the standards laid out by the specification, you ensure that the content you create can be encoded, delivered, and displayed with broadcast hardware and software designed to these standards. This specification sweeps across the entire interactive TV industry and ensures that all the players—hardware manufactures, software companies, broadcasters, cable operators, and content developers—are following the same standards.

The bottom line is that when everybody plays by the same rules, developing and delivering interactive TV content becomes not only easier, but also more cost-effective. A standardized set of interactive content can be handled by any TV receiver built to the specification adopted by the ATVEF. Throughout this book, you will see that the majority of the interactive TV content is based upon the ATVEF's content-creation guidelines.

For more information about designing content to ATVEF guidelines, see the "ATVEF and Content Creation Standards" topic in the "Microsoft TV Programmer's Guide" section of the companion CD.

Use DHTML with Caution

Microsoft TV supports additional functionality that is not included in the specification defined by the ATVEF. For example, Microsoft TV supports a subset of DHTML that enables you to dynamically hide and show elements, move elements, or format elements. Microsoft TV's subset of DHTML, however, should be used with caution, because not all TV receivers support the same level of functionality.

The goal of the "Microsoft TV Design Guide" is to show you how to create browser-independent interactive TV content—meaning content that can be encoded, delivered, and displayed correctly regardless of which TV receiver the content appears in. Accordingly, most of the samples in the book and on the companion CD comply with the content creation standards defined by the ATVEF.

For more information about using DHTML to design interactive TV content, see the "DHTML for Microsoft TV" topic in the "Microsoft TV Programmer's Guide" section of the companion CD.

WHAT'S NEXT

This chapter covered the general design strategies that form the cornerstone of interactive TV design. In Chapter 6, we will look at the screen dimensions for interactive TV design and present and then demonstrate how to layer Web content over a full-screen TV image.

Chapter 6

Layering Web Content over Full-Screen TV

In This Chapter

- Where to Find Sample Content for This Chapter
- Creating an Overlay
- Limitations of Overlays
- About Hiding and Showing DIVs
- Integrating Full-Screen TV into a Web Page
- Making TV Appear in the TV Object
- Creating the Overlay for Lakes & Sons
- How Absolute Positioning Works
- Positioning Overlays for Microsoft TV
- How the Z-Index Property Works
- Transitioning from a Web Page to Full-Screen TV
- Implementing an Order Now Button

An easy way to implement interactive TV content is to layer it over the top of full-screen TV, as is done with the overlays used on TV for sporting events, newscasts, and talk shows. This chapter explains how to create Web-based interactive overlays, how to create links for an overlay to load new content, and how to create a Close button that enables viewers to return to full-screen TV.

WHERE TO FIND SAMPLE CONTENT FOR THIS CHAPTER

To get the most out of this chapter, we recommend that you view the sample files on the *Building Interactive Entertainment and E-Commerce Content for Microsoft TV* companion CD and Web site.

- To view the Lakes & Sons sample content on a set-top box running Microsoft TV or on the Web, go to *http://www.microsoft.com/tv/itvsamples*.

- To view the sample files associated with this chapter, see the "Layering Web Content over Full-Screen TV" topic in the "Microsoft TV Design Guide" section of the companion CD.

- For HTML source files of the generic template, see the "Templates" section of the companion CD.

CREATING AN OVERLAY

Layering Web content over full-screen TV provides an easy and relatively unobtrusive way to introduce interactive TV content to a TV viewer. This approach doesn't work in all cases, but it is ideal for presenting small amounts of information to the viewer in a non-threatening way. For example, you can layer content over a full-screen TV picture to display a recipe during a cooking show. Or you can present content over a full-screen TV picture as the first set of interactive content that the TV viewer sees, as demonstrated in Figures 6.1, 6.2, and 6.3.

To show how an overlay can be used as the initial presentation of interactive content, we have created sample interactive advertising content (Lakes & Sons) that is tied to a fictitious movie called *The Phone Guy*. At the end of the movie, interactive TV links are inserted to enable the viewer to purchase a video copy of the movie. The interactive link at the end of the movie is shown in Figure 6.1. Note that the link could just as easily be tied to a 1-800 commercial at the end of the movie.

Figure 6.1 *An interactive TV link appears at the end of* The Phone Guy.

When the TV viewer clicks the Go Interactive button, the Web content appears on top of the full-screen TV object, as shown in Figure 6.2.

Figure 6.2 *The Web content is placed over the full-screen TV object.*

Viewers can click the Order Now button to load the purchase form shown in Figure 6.3, or click the Close button to return to watching full-screen TV.

Figure 6.3 *The purchase form for Lakes & Sons.*

For instructions on building the purchase form, see Chapter 7, "Layering TV over Web Content."

LIMITATIONS OF OVERLAYS

Layering content over full-screen TV works well for small amounts of information, but it has some limitations. For example, it does not work well for Web content that requires reading from or writing to a database. Due to the nature of the HTTP protocol, when a database is accessed to read or write data, the Web page is refreshed. If the Web page contains full-screen video, the refresh can cause a significant interruption of TV service. For this reason, purchase forms or other types of interactive content that require database connectivity are placed in a frameset. In such cases, one frame of the frameset contains the video object, while another frame contains the content that accesses the database. In this way, the page with the video object remains static while the other page is refreshed.

For details on creating a frameset with the video object on one page and a form on the other, see Chapter 7, "Layering TV over Web Content."

About Transparency

Microsoft TV supports transparency in two ways. It supports a proprietary transparency property for the TABLE element, and it supports the transparent attribute of the cascading style sheets (CSS) background-color property.

The Transparency Property of the TABLE Element

Microsoft TV supports a proprietary transparency property for the TABLE element. This property allows you to set a range from 0 (fully opaque) to 100 (fully transparent). However, because transparency is limited to Microsoft TV, it is not used in the samples in this book. If you use the transparency property in your interactive TV content, be aware that it may only work on Microsoft TV. The following example shows how to specify transparency for a table.

```
<table bgcolor="#332949" transparency="50">
```

The Transparent Attribute of the CSS Background-Color Property

Microsoft TV supports the transparent attribute for the CSS background-color property, so you can specify a transparent background color for a document or table. However, this transparent attribute does not support a percentage or number value, so you cannot adjust the degree of transparency. The syntax for specifying a transparent background-color property for a document is:

```
<body style="background-color:transparent">
```

Note that the transparent attribute of the CSS background-color property is also not used in the sample content in this book.

ABOUT HIDING AND SHOWING DIVS

Dynamic HTML (DHTML) can be useful for creating divisions that will hide and show small amounts of information, such as overlays over full-screen TV. Nevertheless, DHTML should be used with caution, because it is not guaranteed to work on all TV receivers. What's more, you will most likely need to write two sets of code to handle the object model differences between Netscape-based TV receivers and Microsoft TV–based receivers. That is because Netscape references layers; Microsoft TV references divisions. The samples in this book do not use DHTML to hide and show divisions. Rather, overlays are presented for initial interactive content, and a frameset solution is used as a method of dynamically swapping content.

INTEGRATING FULL-SCREEN TV INTO A WEB PAGE

The Advanced Television Enhancement Forum (ATVEF) identifies several ways to implement a full-screen TV object into a Web page using the "tv:" attribute as a substitute for a Uniform Resource Locator (URL). According to the ATVEF, the "tv:" URL may be used whenever an image URL is also appropriate. One of the easiest ways to implement full-screen TV for a Web page is to specify the "tv:" attribute for the background property of a <body> tag, as shown in the following example:

```
<body background="tv:">
```

A more robust way to implement full-screen TV is to use the <object> tag. This strategy is advisable because some TV receivers will not support the <body background="tv:"> tag. The following code snippet shows the best way to implement full-screen TV for Microsoft TV:

```
<div style="position:absolute;top:0;left:0;z-index:-1">
<object data="tv:" height=100% width=100%> </object>
</div>
```

Next, we'll examine some of the details of the full-screen implementation shown in the sample code, including using absolute positioning and the z-index property.

MAKING TV APPEAR IN THE TV OBJECT

A common experience among those who are new to working with the TV object is to create a Web page with the TV object in it and then open the page in a standard personal computer–based browser, only to discover that the TV object looks like an image with a broken link. To view video in a TV object, you must view the Web page with a TV receiver built to ATVEF standards. Also, you must post the page to an external Web server. In addition, for Microsoft TV, you must force the page into TV mode. To do this, you can create a page with a link that forces the page into TV mode using the view="tv" attribute for the link. This page represents an interactive TV link sent during a television show or commercial. For the samples in this book, a viewset.html page, as shown in Figure 6.4, is used to open the lakes_01.html file.

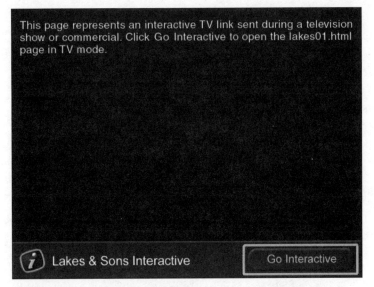

Figure 6.4 *The viewset.html page for the Lakes & Sons content.*

The following code shows how the viewset.html page is constructed. The anchor that sets TV mode is shown in bold.

```
<html>
<!-- This page sets the view mode to TV
and opens the lakes01.html page. -->
<head>
<title>Lakes & Sons Interactive</title>
</head>

<body bgcolor="#000000" text="#C0C0C0">
<div style="position:absolute;top:25;left:8">
<font size=3 face="Arial,Verdana,Helvetica">
This page represents an interactive TV link sent during a television show
or commercial. Click Go Interactive to open the lakes01.html page
in TV mode.</font>
</div>

<div style="position:absolute;top:322;left:0">
<table background="images/UI_544bac.jpg" border=0 width="544" height="57">
<tr>
<td width=355> </td>
```

(continued)

```
<td valign="center"><a href="lakes01.html" view="tv"><img src="images/
go.gif" width=172 height=37 alt="" border="0"></a></td>
</tr>
</table>
</div>
</body>
</html>
```

You can find the viewset.html file in the Lakes template folder on the companion CD. The viewset.html page, pictured earlier in Figure 6.4, features a Go Interactive button that, when clicked, forces the TV receiver into TV mode, so the TV picture appears in the TV object. Opening lakes01.html with the viewset.html file also sets TV mode for the Microsoft WebTV 2.0 Viewer.

CREATING THE OVERLAY FOR LAKES & SONS

To put the full-screen TV object solution in context, we'll examine how it is used with the Lakes & Sons interactive content described earlier. As is indicated in bold in the following code, the TV object is contained within DIV tags. The division is positioned absolutely at the upper left corner of the TV screen. The z-index, which controls the layering of elements in a document, is set to a negative number to ensure that the division containing the TV object is the bottom-most division in the document. The DIV that contains the overlay is then placed over the TV object.

```
<html>
<head>
<title>Lakes & Sons Interactive</title>
</head>

<body>

<!-- This div contains the TV object. -->
<div style="position:absolute;top:0;left:0;z-index:-1">
<object data="tv:" height=100% width=100%>
</object>
</div>

<!-- This div contains the overlay. -->
<div style="position:absolute;top:8;left:0">
<img src="images/imap.jpg" width=189 height=322 alt="" border="0"
usemap="#ordermap">
</div>

<!-- These are the links for the image map. The top link
returns to full-screen TV. The bottom link opens the order form
frameset. -->
```

```
<map name="ordermap">
<area shape="rect" coords="61,293,124,315" href="tv:">
<area shape="rect" coords="25,217,156,270" href="lakes_main.html">
</map>

</body>
</html>
```

HOW ABSOLUTE POSITIONING WORKS

Traditionally, most elements in HTML are positioned relative to previous elements in the flow of the document. However, Microsoft TV supports CSS positioning, so elements can be positioned on a fixed plane separate from the document's flow or offset from the traditional position in a document. CSS positioning allows elements to overlap and provides Web authors with more control over the layout than was previously possible.

The CSS position property takes one of three values: *static*, *absolute*, and *relative*. Static positioning, by default, has no affect on the traditional layout of the HTML element. Relative positioning is used to offset an element from its normal position in the flow, and is rarely used in our samples. Absolute positioning, used throughout the samples, is used to specify a fixed location for the element outside the flow of the document. Absolute positioning works well for Microsoft TV because the TV window has a fixed size of 560-by-420 pixels.

Absolutely positioned elements are positioned with respect to some containing element's coordinate system. The upper left corner of the document (top 0, left 0) defines the coordinate system for all absolutely positioned elements. For example, the following code places the top left corner of a division at a top coordinate of 20 pixels and a left coordinate of 20 pixels:

```
<body>
<div style="position:absolute;top:20;left:20;">
This is division 1.
</div>
</body>
```

Whenever an element is relatively or absolutely positioned, a new coordinate system is defined for all the elements it contains. For example, let's embed a division within the division shown above.

```
<body>
<div style="position:absolute;top:20;left:20;">
<div style="position:absolute;top:20;left:20;">
This is division 2.
</div>
</div>
</body>
```

Now the embedded division is positioned relative to its parent container, which is the first division (coordinates 20, 20). As a result, the embedded division is positioned at top and left coordinates of 40, 40, respectively.

For more information about using CSS properties, see Chapter 8, "Formatting Microsoft TV Content with Styles and Style Sheets," and Chapter 23, "CSS Level 1 Support for Microsoft TV."

POSITIONING OVERLAYS FOR MICROSOFT TV

Microsoft TV provides a slight complication in regard to absolute positioning, because it has a fixed left margin of 8 pixels. However, the top margin is not fixed and can be set at 0. Therefore, the way to create a consistent top and left border for an overlay is to set the top coordinate to 8 and the left coordinate to 0, as shown in bold in the following code snippet:

```
<!-- This div contains the overlay. -->
<div style="position:absolute;top:8;left:0">
<img src="images/imap.jpg" width=189 height=322 alt="" border="0"
usemap="#ordermap">
</div>
```

This solution positions the image 8 pixels from the top border and 8 pixels from the left border, as shown in Figure 6.5.

Figure 6.5 *The Lakes & Sons overlay.*

IMPORTANT   With regard to Microsoft TV, absolute positioning is supported only for the <div> tag, so you will see it used with that tag throughout this book.

HOW THE Z-INDEX PROPERTY WORKS

The z-index property defines the graphical z-order, or overlapping of elements in relation to other elements. By default, all elements that define a coordinate system, including the <body> element, are positioned with a z-index of 0. Other elements can be positioned behind the text by having a negative z-index value. Elements whose z-index values are not specified are implicitly assigned z-index values according to their source order in a document. Therefore, an element that is positioned later in a document is displayed above any elements positioned earlier. Elements positioned later in a document have a higher z-index value than those elements positioned earlier in a document. As shown in bold in the following code snippet, the z-index of the DIV containing the TV object for the Lakes & Sons overlay is positioned at the bottom layer of the document by setting its z-index value to a negative number:

```
<!-- This div contains the TV object. -->
<div style="position:absolute;top:0;left:0;z-index:-1">
<object data="tv:" height=100% width=100%>
</object>
</div>
```

TRANSITIONING FROM A WEB PAGE TO FULL-SCREEN TV

You can use "tv:" as the href of an anchor tag to return to full-screen TV. For example:

```
<a href="tv:">click here to return to TV</a>
```

For the Lakes & Sons content, an image map is defined for the overlay, and the <a href="tv:"> anchor is defined for the Close button portion of the image, as shown in bold in the following code example:

```
<!-- This div contains the overlay. -->
<div style="position:absolute;top:8;left:0">
<img src="images/imap.jpg" width=189 height=322 alt="" border="0"
usemap="#ordermap">
</div>

<!-- These are the links for the image map. The top link
returns to full-screen TV. The bottom link opens the order form
frameset. -->
<map name="ordermap">
```

(continued)

```
<area shape="rect" coords="61,293,124,315" href="tv:">
<area shape="rect" coords="25,217,156,270" href="lakes_main.html">
</map>
```

IMPLEMENTING AN ORDER NOW BUTTON

The Order Now button is a link on the overlay image. This button is linked to the lakes_main.html page, a frameset-based set of content that enables the viewer to order a video of *The Phone Guy* online while watching TV. The code for the Order Now button is shown in bold in the following code:

```
<!-- This div contains the overlay. -->
<div style="position:absolute;top:8;left:0">
<img src="images/imap.jpg" width=189 height=322 alt="" border="0"
usemap="#ordermap">
</div>

<!-- These are the links for the image map. The top link
returns to full-screen TV. The bottom link opens the order form. -->
<!--frameset-->
<map name="ordermap">
<area shape="rect" coords="61,293,124,315" href="tv:">
<area shape="rect" coords="25,217,156,270" href="lakes_main.html">
</map>
```

For more information about creating a frameset for interactive TV content, see Chapter 7, "Layering TV over Web Content."

WHAT'S NEXT

This chapter explained how to create a full-screen TV object on a Web page and how to overlay content on the TV object. It also explained how to position an overlay on the screen, how to set its z-index property, and how to transition from a Web page to full-screen TV. In the following chapter, we will look at how to layer the TV object over Web content. We will also look at how to create frames-based interactive TV content that serves as the foundation for e-commerce applications.

Chapter 7

Layering TV over Web Content

In This Chapter

■ Where to Find Sample Content for This Chapter

■ Using a Frameset to Lay Out Interactive TV Content

■ Designing an HTML TV Page

■ Formatting Text with Style Sheets

■ Creating an Interactive E-Commerce Page

This chapter provides the fundamental concepts for designing robust, extensible interactive TV content that can be easily optimized for database access. To create a useful and entertaining interactive TV experience, you must design content that changes dynamically or enables the TV viewer to submit information to a database. At the same time, however, you must provide a continual TV picture, free from the interruption that can be caused by a page refresh. This chapter shows how to design content to meet these demands by using a frames-based solution.

The previous chapter demonstrated how to layer interactive Web content over a full-screen TV object. This chapter provides a natural progression from the previous chapter by describing how to layer the TV object over Web content. As demonstrated in this chapter, one frame contains the TV object, which remains constant on the screen, while the other frame serves as a container for content that can be swapped out, updated, and refreshed.

In this chapter, we offer a holistic approach to adding the TV object to a Web page. Rather than simply describing how to add the TV control to a Web page, we present a complete solution for designing useful, dynamic content that includes the TV object. We will use the Lakes & Sons sample content to demonstrate the following fundamental concepts:

- How to create a frameset to divide interactive content into static and dynamic frames

- How to design content to fit within the dimensions of Microsoft TV

- How to create the TV object, including setting its dimensions and positioning it on the page

- How to create a border for the TV object

- How to implement a Full Screen button to enable the TV viewer to transition from interactive content back to full-screen TV

- How to build a form for Microsoft TV

WHERE TO FIND SAMPLE CONTENT FOR THIS CHAPTER

To get the most out of the material in this chapter, we recommend that you view the sample files on the *Building Interactive Entertainment and E-Commerce Content for Microsoft TV* companion CD and Web site.

- To view the Lakes & Sons content on a set-top box running Microsoft TV or on the Web, go to *http://www.microsoft.com/tv/itvsamples*.

- To view the sample files associated with this chapter, see the "Layering TV over Web Content" topic in the "Microsoft TV Design Guide" section of the companion CD.

- To view HTML source files of the generic template, see the "Templates" section of the companion CD.

An Overview of the Lakes & Sons Content

The Lakes & Sons content covered in this chapter provides a good example of how to integrate the TV object into Web content. Figure 7.1 shows the Lakes & Sons interactive content when it is loaded and the TV viewer clicks the Order Now button on the overlay, as described in Chapter 6, "Layering Web Content over Full-Screen TV."

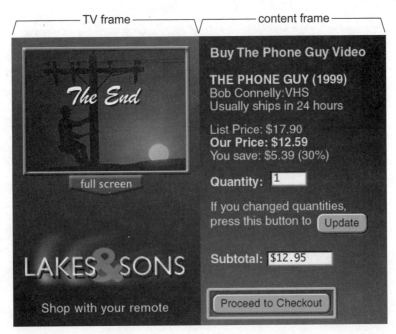

Figure 7.1 *Lakes & Sons interactive content.*

As shown in Figure 7.1, the Lakes & Sons content is divided into two frames. The left frame, called the tv frame, contains a Web page that holds the TV object. The page is always visible in the frameset. The right frame, called the content frame, dynamically swaps out pages as necessary to enable the TV viewer to complete the purchase of *The Phone Guy* video online.

Designed for Microsoft TV screen dimensions, Microsoft TV provides a design space of 560 pixels in width by 420 pixels in height, so the Lakes & Sons content is specifically designed to fit neatly within this space. As you can see in Figure 7.2, the overall width of the critical content for the page is 528 pixels, leaving a 16-pixel border around the content.

Keep in mind that the 528-pixel content width is not cast in stone. Rather, it is an arbitrary width, which we found to work quite nicely for presenting interactive TV content. Microsoft TV does impose a left border of 8 pixels by default, but this can be easily overridden, as we will explain later in this chapter.

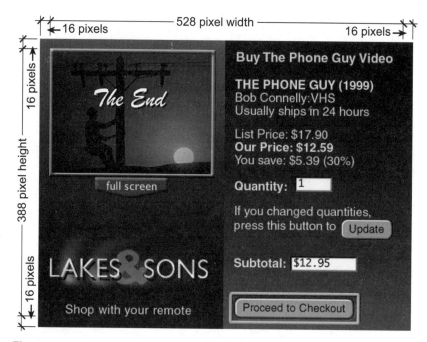

Figure 7.2 *Content dimensions for Lakes & Sons interactive.*

USING A FRAMESET TO LAY OUT INTERACTIVE TV CONTENT

The most flexible and robust way to design interactive TV content is with a frameset and frames. Frames are often disparaged in developer circles and do not exactly represent cutting-edge technology, but for interactive TV, a frames-based solution offers several distinct advantages:

■ With frames, you can continually display the TV object in one frame, while changing content in another.

■ Using a frameset and frames complies with the specification defined by the Advanced Television Enhancement Forum (ATVEF). At first glance, it may seem like a good idea to use <div> tags and JavaScript as a dynamic way of swapping out content. However, the Document Object Model (DOM) specified by the ATVEF is essentially JavaScript 1.1, which does not support hiding or showing divisions and layers.

■ With frames, you can divide interactive TV content into small pages that not only download quickly, but also are easy to manage and update. By comparison, a page that uses dynamic HTML (DHTML) to hide and show divisions is typically laden with JavaScript and requires programming expertise to maintain. Further, these pages quickly become large and take longer to download.

■ With frames, you can swap out content while maintaining the same URL for the page, thus avoiding the "Getting Enhancement" screen you see when Microsoft TV fetches a new page from a new URL.

■ With frames, you can swap out pages in the frameset by sending broadcast triggers to the main page of the frameset.

Avoiding Iframes

Iframes are inline frames that you can specify for a Web page. At first glance, iframes appear to be the perfect solution for designing interactive TV content. In fact, the authors of this book tried using iframes to lay out content. Iframes work great as a means of presenting basic information, but proved to be unreliable for handling JavaScript in Microsoft TV. Consequently, we do not use iframes for any samples in this book. As a general rule, use iframes only to present HTML content that does not require JavaScript.

Designing the Frameset for Lakes & Sons

The Lakes & Sons content, as shown previously in this chapter, is a frameset-based solution that consists of the following Web pages:

■ **lakes_main.html**—defines the frameset and the frames for the content.

■ **lakes_tv.html**—is the Web page that appears in the left-hand frame, called the tv frame, of the frameset. This page, which is always visible in the left-hand frame, contains the TV object.

■ **lakes_content.***x*—are the content pages for the right-hand frame, called the content frame, of the frameset. Web pages in the content frame are dynamically swapped out based on hyperlink selections or button clicks made by the TV viewer.

The lakes_main.html page that defines the frameset is relatively easy to create, as shown in the following code sample:

```
<html>
<head>
<title>Lakes & Sons</title>
</head>
<!-- JavaScript removes the gray borders from around a frame for WebTV. -->
<script type="text/JavaScript">
if (navigator.appVersion.indexOf("WebTV") != -1)
{
    document.write("<body hspace=0 vspace=0>");
}
</script>

<frameset cols="281,*" framespacing=0 frameborder=0
hspace=0 vspace=0 marginwidth=0 marginheight=0>
<frame name="tv" src="lakes_tv.html" marginwidth=0 marginheight=0
scrolling=0 noresize frameborders="no">
<frame name ="content" src="lakes_content1.html" marginwidth=0
marginheight=0 scrolling="no" noresize frameborders ="no">
</frameset>
</html>
```

Using JavaScript to Remove the Frameset Borders

Microsoft TV is essentially the WebTV Plus browser built on top of the Microsoft Windows CE operating system. One of the anomalies of WebTV is that it creates a gray border around framesets that cannot be removed by setting the FRAMEBORDER attribute at 0. To remove the border, you must use JavaScript, placing the border at the top of the HTML page. This code, shown in the following code sample, may seem slightly unorthodox, but it does remove the unwanted border.

```
<!-- JavaScript removes the gray borders from around a frame for WebTV. -->
<script type="text/JavaScript">
if (navigator.appVersion.indexOf("WebTV") != -1)
{
    document.write("<body hspace=0 vspace=0>");
}
</script>
```

Creating the TV and Content Frames

As noted earlier, the frameset for lakes_main.html is divided into two frames, the tv frame on the left and the content frame on the right. The tv frame is given a fixed width of 281 pixels, while the content frame is set to *, which specifies that the frame expand or contract to fill the remaining space on the TV screen. With this technique, demonstrated in bold in the following sample code, the right-hand frame resizes to accommodate the window in which it is loaded. Regardless of the window width, however, the width of the content in the frameset remains a fixed 528 pixels and the height remains 388 pixels.

```
<frameset cols="281,*" framespacing=0 frameborder=0
hspace=0 vspace=0 marginwidth=0 marginheight=0>
<frame name="tv" src="lakes_tv.html" marginwidth=0 marginheight=0
scrolling=0 noresize frameborders="no">
<frame name ="content" src="lakes_content1.html" marginwidth=0
marginheight=0 scrolling="no" noresize frameborders ="no">
</frameset>
```

Creating Non-Scrollable Content

At the time of this writing, Microsoft TV does not support scrollable pages in TV mode. As a result, if content extends beyond the visible boundary of the TV screen, that content becomes inaccessible to the TV viewer. In addition, Microsoft TV does not support scrollable frames. Instead, it converts frames into tables, so frames that require scrolling are often transformed into long vertical tables. When designing content for Microsoft TV, avoid creating scrollable content, if possible.

DESIGNING AN HTML TV PAGE

The TV page is the page that contains the TV control and, in many cases, navigational elements. Figure 7.3 shows the TV page, named lakes_ tv.html, created for the Lakes & Sons interactive content.

Figure 7.3 *The lakes_tv.html page.*

To help you understand how the lakes_tv.html page is created, we will first show you the HTML code, and then dissect it to explain how various components of the page are constructed. Here, then, is the HTML code for the lakes_tv.html page:

```html
<html>
<head>
<!--links the cssTV.css stylesheet for Microsoft TV-->
<link rel="stylesheet" type="text/css" href="cssTV.css">

<!--links the cssPC.css stylesheet for design-time purposes-->
<script type="text/JavaScript">
var bName = navigator.appName
if (bName == "Microsoft Internet Explorer")
{
    document.write("<link rel='stylesheet' type='text/css'
        href='cssPC.css'>");
}
</script>
<title>Lakes & Son Interactive</title>
</head>

<body background="images/tile1.gif">

<div style="position:absolute;top:14;left:14">
<img src="images/border.gif" width=250 height=187 alt="" border="0">
</div>

<div style="position:absolute;top:20;left:20">
<object data="tv:" width=238
height=179 alt="" border="0"></object>
</div>

<div style="position:absolute;top:205;left:80">
<a href="tv:"><img src="images2/full.gif" width=117 height=33 alt=""
border="0"></a>
</div>

<div style="position:absolute;top:271;left:14">
<img src="images/logo.jpg" width=256 height=89 alt="" border="0">
</div>

<div style="position:absolute;top:382;left:38">
<font class="clsDescription">Shop with your remote</font>
</div>

</body>
</html>
```

Creating a Gradient for the Page

Microsoft TV supports a GRADIENT attribute for tables, but we did not use it because it is a proprietary tag and may not work on other interactive TV receivers. Rather, we achieved the same effect by using a small graphic that tiles on the page. This creates a pleasing effect without using a lot of graphic overhead. The following code shows how we added the graphic to the page:

```
<body background="images/tile1.gif">
```

Setting the Margins for the Page

For the lakes_main.html page, the page that defines the frameset, we used JavaScript to remove any borders around the frameset. We also set the marginheight and marginwidth of the tv and content frames to 0. Absolute positioning was then used to place elements on the page. As you can see in the following code sample, absolute positioning places the top-left corner of the TV border at 14 pixels from the top of the document, and 14 pixels from the right border of the document. Strangely, the JavaScript solution that removes the gray borders leaves a few pixels behind, so the total package gives you a right and top border of 16 pixels each. Because the overall content width of Lakes & Sons is 528 pixels and the height is 388 pixels, an overall 16-pixel margin is created for the critical content on the page.

```
<div style="position:absolute;top:14;left:14">
<img src="images/border.gif" width=250 height=187 alt="" border="0">
</div>
```

Creating the Border for the TV Object

Although it is not absolutely necessary, positioning a border image behind the TV object, as shown previously in Figure 7.3, adds professional polish to the page and makes it look more TV-friendly. The border for the TV object is a .gif image with dimensions following the 4:3 aspect ratio required for National Television Standards Committee (NTSC) broadcasts. Because the width of the border is 250 pixels, you multiply by .75 to get the 187-pixel height of the border. Also notice that the border is the first element in the HTML document, thus positioning it at the bottom layer of the page. By specifying the DIV that contains the TV object after the DIV that contains the border, as shown in the following sample code, the TV object is effectively layered over the top of the border. The effect is that the border and the TV object appear to be a single element on the page, as shown in Figure 7.4.

```
<div style="position:absolute;top:14;left:14">
<img src="images/border.gif" width=250 height=187 alt="" border="0">
</div>
```

(continued)

```
<div style="position:absolute;top:20;left:20">
<object data="tv:" width=238
height=179 alt="" border="0"></object>
</div>
```

Figure 7.4 *The border.gif image is positioned on the page.*

Creating the TV Object

The specification defined by the ATVEF offers a variety of easy ways to add the TV object to a Web page. According to the specification, you add a broadcast television channel to a Web page by specifying "tv:" as an attribute of the <object>, <img>, <body>, <frameset>, <a>, <div>, or <table> tag. For example, the following statement adds the TV object to a Web page:

```
<img src="tv:" width="200" height="150" border="0">
```

Although the ATVEF offers a variety of methods for adding the TV object to a Web page, the most reliable method at present is to use the following syntax:

```
<object data="tv:" height=x width=x>
```

The above syntax works across the widest variety of interactive TV receivers, so it is no coincidence that it is used throughout this book.

Positioning the TV Object

The TV object for the lakes_tv.html page meets two important requirements:

■   It fits seamlessly over the border image.

■   Its width and height conform to a 4-to-3 aspect ratio.

To achieve the above criteria, the TV object is positioned over the border.gif image. However, it is offset 6 pixels to the right and 6 pixels down, as shown in the following code sample. The DIV containing the border, shown in bold, is positioned at (top:14;left:14), while the DIV containing the TV object is positioned at (top:20;left:20). This offset enables the 6-pixel border image to be visible.

```
<div style="position:absolute;top:14;left:14">
<img src="images/border.gif" width=250 height=187 alt="" border="0">
</div>

<div style="position:absolute;top:20;left:20">
<object data="tv:" width=238
height=179 alt="" border="0"></object>
</div>
```

Specifying the Dimensions for the TV Object

The dimensions for the TV object are calculated by subtracting the width of the borders in border.gif from the height and width of the border image. For example, the border.gif image is 250 pixels in width by 187 pixels in height. To determine the width of the TV object, take the combined width of the right and left borders, which is 12, and subtract from the overall width of the image, giving you a TV object width of 238 pixels. The same procedure is used to determine the height of the TV object.

```
<div style="position:absolute;top:20;left:20">
<object data="tv:" width=238
height=179 alt="" border="0"></object>
</div>
```

Making TV Appear in the TV Object

As we noted in Chapter 6, a viewset.html page was used to force the Lakes & Sons content into TV mode. For the Lakes & Sons content, the viewset.html page calls the lakes01.html page. This overlay page contains an Order Now button, which when clicked opens the lakes_main.html page. Because the TV receiver is already in TV mode, video appears in the TV object of the lakes_tv.html page. You can find the viewset.html file in the Lakes template folder on the companion CD.

Implementing the Full Screen Button

When designing interactive TV content, it is important to provide users with an easy way to switch from viewing full-screen TV to viewing interactive TV content. For the Lakes & Sons content, this is accomplished with a hyperlink labeled Full Screen, as shown in Figure 7.5. Usability studies have shown that the Full Screen or TV Only terminology is easily understood by users.

Figure 7.5 *The Full Screen button on the lakes_tv.html page.*

When clicked, the Full Screen button switches the TV viewer from a page that contains Web content and a reduced-sized TV object to a page that shows a full-screen television picture, as shown in Figure 7.6.

Figure 7.6 *The television screen in full-screen mode.*

Positioning the Full Screen Button

In almost all of the interactive TV pages covered in this book, the Full Screen button abuts the border of the TV object, creating the illusion that it is actually part of the border. To position the Full Screen button, <div> tags and absolute positioning are used, as shown in the following sample code. The border is specified in the first division; the TV control is contained and positioned in the second division; and the Full Screen button is positioned in the third division, so that its top boundary abuts the bottom of the border.

```
<div style="position:absolute;top:14;left:14">
<img src="images/border.gif" width=250 height=187 alt="" border="0">
</div>

<div style="position:absolute;top:20;left:20">
<object data="tv:" width=238
height=179 alt="" border="0"></object>
</div>

<div style="position:absolute;top:205;left:80">
<a href="tv:"><img src="images/full.gif"
width=117 height=33 alt="" border="0"></a>
</div>
```

Defining the Anchor for the Full Screen Button

To enable TV viewers to return to full-screen TV, you must specify "tv:" as the HREF attribute of the <a> tag, as shown in the following sample code:

```
<div style="position:absolute;top:205;left:80">
<a href="tv:"><img src="images/full.gif"
width=117 height=33 alt="" border="0"></a>
</div>
```

Positioning Content at the Bottom of the Page

Because Microsoft TV uses a fixed window size, it is important to position elements to create a pleasing bottom border for the page. For the lakes_tv.html page, the tagline "Shop with your remote" is positioned on the page to line up with the Proceed to Checkout button on the lakes_content1.html page, as shown in Figure 7.7.

Figure 7.7 *"Shop with your remote" is aligned with the Proceed to Checkout button.*

Most of the samples in this book feature a 16-pixel margin at the bottom of the page. As shown in Figure 7.2, the bottom border of the Proceed to Checkout button is located 404 pixels down the page. An easy way to determine where the 404-pixel point is on the page, especially for text or form elements for which you cannot set the exact height dimensions, is to position a graphic image at the (top:404 pixel) coordinate. You can then place the bottom portion of the text or form element in the same position as the top border of the image. The following code shows how the logo and "Shop with your remote" tagline are positioned on the lakes_tv.html page.

```
<div style="position:absolute;top:271;left:14">
<img src="images/logo.jpg" width=256 height=89 alt="" border="0">
</div>

<div style="position:absolute;top:382;left:38">
<font class="clsDescription">Shop with your remote</font>
</div>
```

FORMATTING TEXT WITH STYLE SHEETS

The lakes_tv.html page uses an alternate style sheet solution to handle the differences in text sizes between the computer and the TV. A TV style sheet, cssTV.css, is used for TV display, and a personal computer style sheet, cssPC.css, is used for computer display. The cssPC.css style sheet helps you create interactive TV content on the computer because it approximates the size of the fonts used by Microsoft TV. In this way, what you see on the computer looks very similar to what you will see on Microsoft TV. The following code shows how this alternate style sheet solution is implemented. By default, the cssTV.css style sheet is attached to the document; however, if the Microsoft Internet Explorer appName is detected, the cssPC.css style sheet is attached. This solution is not required, but its use makes it easier to design interactive TV content on a computer.

```
<!--links the cssTV.css stylesheet for WebTV-->
<link rel="stylesheet" TYPE="text/css" href="cssTV.css">

<!--links the cssPC.css stylesheet for design-time purposes-->
<script type="text/JavaScript">
var bName = navigator.appName
if (bName == "Microsoft Internet Explorer")
{
    document.write("<link rel='stylesheet' type='text/css'
        href='cssPC.css'>");
}
</script>
```

Using the cssTV.css and the cssPC.css Style Sheets

As you can see, the cssTV.css style sheet is very simple and specifies font sizes 18 point and above for TV. The 18-point font size is the optimum point size for TV.

```
A {font-family:Arial,Verdana,Helvetica;font-size:18pt;
color:#e6ba72;text-decoration:none}

.clsTitle {font-family:Arial,Verdana,Helvetica;
font-size:20pt;color:#dacefb;font-weight:bold}

.clsDescription {font-family:Arial,Verdana,Helvetica;
font-size:18pt;color:#e6ba72}
```

The cssPC.css style sheet specifies smaller font sizes. These font sizes for the personal computer map quite closely to the font sizes in the cssTV.css style sheet for TV.

```
A {font-family:Arial,Verdana,Helvetica;font-size:13pt;color:#e6ba72;
text-decoration:none}

.clsTitle {font-family:Arial,Verdana,Helvetica;font-size:14pt;
font-weight:bold;color:#dacefb}

.clsDescription {font-family:Arial,Verdana,Helvetica;
font-size:13pt;color:#e6ba72;font-weight:bold}
```

Assigning Classes to Format Text

Microsoft TV supports the use of the class attribute for the FONT element, so you can use the class attribute to format text. Classes give you great flexibility because the classes format text differently, based on whether they are shown on TV or on the computer. For example, we applied the .clsDescription class to the "Shop with your remote" text sample. When we designed the sample on the computer with a Microsoft Internet Explorer browser, we used the cssPC.css style sheet. When shown on Microsoft TV, however, the .clsDescription class causes the cssTV.css style sheet to be substituted for the css.PC.css class. The following shows how the clsDescription class is applied to the "Shop with your remote" text:

```
<div style="position:absolute;top:382;left:38">
<font class="clsDescription">Shop with your remote</font>
</div>
```

As shown in the preceding code snippet, the same class is used to format the text, but the class formats the text differently for the computer than for Microsoft TV. Using dual style sheets and classes makes it easy to design on the computer because what you see on the computer very closely approximates what you will see on TV.

If this solution is not completely sinking in right now, don't worry. The application of styles, classes, and dual style sheets is covered in more detail in Chapter 8, "Formatting Microsoft TV Content with Styles and Style Sheets," and Chapter 10, "Creating Text for Microsoft TV."

CREATING AN INTERACTIVE E-COMMERCE PAGE

The lakes_content1.html page is designed to be the first of a series of e-commerce pages that enable the user to purchase a video online. Note that lakes_content1.html is the only page currently designed for this sample. The lakes_content1.html page is shown in Figure 7.8.

To help you understand how the lakes_content1.html page is constructed, we will first show you the HTML code and then dissect it piece-by-piece to explain how various components of the page were created. Here, then, is the HTML code for the lakes_content1.html page:

```
<html>
<head>
<!--links the cssTV.css stylesheet for WebTV-->
<link rel="stylesheet" type="text/css" href="cssTV.css">

<!--links the cssPC.css stylesheet for design-time purposes-->
<script type="text/JavaScript">
var bName = navigator.appName
```

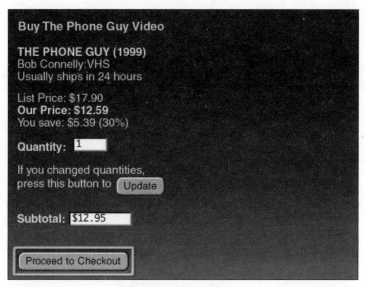

Figure 7.8 *The lakes_content1.html page.*

```
if (bName == "Microsoft Internet Explorer")
{
    document.write("<link rel='stylesheet' type='text/css'
        href='cssPC.css'>");
}
</script>
<title>Lakes and Sons Content</title>
</head>

<body background="images/tile2.gif" width=13 height=420 alt="" border="0">

<div style="position:absolute;top:14;left:14;">
<font class="clsTitle">Buy The Phone Guy Video</font>
</div>

<div style="position:absolute;top:54;left:14;">
<font class="clsDescription">THE PHONE GUY(1999)<br>
Bob Connelly:VHS<br>
Usually ships in 24 hours</font>
</div>

<div style="position:absolute;top:124;left:14;">
<font color="#F0CA8F" face="Arial,Verdana,,Helvetica" size=3>List
Price: $17.90<br>
<b>Our Price: $12.59</b><br>
```

(continued)

```
You save: $5.31 (30%)</font>
</div>

<form>
<div style="position:absolute;top:194;left:14;">
<font color="#F0CA8F" face="Arial,Verdana,,Helvetica"
size=3><b>Quantity:  </b><input type="Text" value="1"
name="ListPrice" size="5" maxlength="5"></font>
</div>

<div style="position:absolute;top:234;left:14;">
<font color="#F0CA8F" face="Arial,Verdana,,Helvetica" size=3>If you
changed quantities,<br> press this button to</font>
<input type="Button" name="subTotal" value="Update" align="absmiddle">
</div>

<div style="position:absolute;top:310;left:14;">
<font color="#F0CA8F" face="Arial,Verdana,,Helvetica"
size=3><b>Subtotal: </b><input type="text" value="$12.59"
name="Subtotal" size="10" maxlength="10" readonly></font>
</div>

<div style="position:absolute;top:373;left:14;">
<input type="Button" name="btnProceed" value="Proceed to Checkout"
align="absmiddle" selected>
</div>

</form>

</body>
</html>
```

Setting a Background Tile

A simple .gif image, tile2.gif, is used to set the background gradient for the page. The tile2.gif image, 13 pixels wide by 420 pixels high, is a long thin image that tiles on the screen to create the gradient effect. The following code snippet shows how the gradient is created:

```
<body background="images/tile2.gif" width=13 height=420 alt="" border="0">
```

> **NOTE**   Microsoft TV provides a GRADIENT attribute, but it is not used in the book samples because it is a proprietary attribute and is not supported by other interactive TV receivers.

Positioning Content on the Page

As shown in bold in the following code snippet, margins on the lakes_content1.html page are set much the same as the margins for the lakes_tv.html page. The topmost content is positioned 14 pixels from the top of the document, while the left margin is defined as 14 pixels to create a consistent border for the frameset. Once again, notice that absolute positioning is used to place elements on the page.

```
<div style="position:absolute;top:14;left:14;">
<font class="clsTitle">Buy The Phone Guy Video</font>
</div>

<div style="position:absolute;top:54;left:14;">
<font class="clsDescription">THE PHONE GUY(1999)<br>
Bob Connelly:VHS<br>
Usually ships in 24 hours</font>
</div>
```

Using the Selected Attribute

Microsoft TV provides a special selected attribute that allows you to set the focus on a link or on a form element. In this case, the selected attribute is applied to the btnProceed button so the focus is on the button when the lakes_content1.html page is first loaded.

```
<input type="Button" name="btnProceed" value="Proceed to Checkout"
align="absmiddle" selected>
```

Creating the Form

The form for the lakes_content1.html page is relatively simple and follows general HTML guidelines for creating a form. The following code snippet shows how the form is created:

```
<form>
<div style="position:absolute;top:194;left:14;">
<font color="#F0CA8F" face="Arial,Verdana,,Helvetica" size=3>
<b>Quantity:  </b>
<input type="Text" value="1" name="ListPrice" size="5" maxlength="5">
</font>
</div>

<div style="position:absolute;top:234;left:14;">
<font color="#F0CA8F" face="Arial,Verdana,,Helvetica" size=3>
If you changed quantities,<br> press this button to</font>
<input type="Button" name="subTotal" value="Update" align="absmiddle">
</div>
```

(continued)

```
<div style="position:absolute;top:310;left:14;">
<font color="#F0CA8F" face="Arial,Verdana,,Helvetica" SIZE=3>
<b>Subtotal: </b>
<input type="Text" value="$12.59" name="Subtotal" size="10"
maxlength="10" readonly></font>
</div>

<div style="position:absolute;top:373;left:14;">
<input type="Button" name="btnProceed" value="Proceed to Checkout"
align="absmiddle" selected>
</div>

</form>
```

We will go into the details of creating a form and linking it to a database in Chapter 17, "Creating Forms for Microsoft TV Content."

WHAT'S NEXT

This chapter demonstrated how to create a frames-based interactive TV solution, using one frame to host the page with the TV object and another frame to host the form pages for e-commerce. This chapter also explained how to layer TV over Web content and how to implement a Full Screen button to transition from interactive Web content to full-screen TV. In the following chapter, we will take an in-depth look at how to use styles and style sheets to format interactive TV content for Microsoft TV.

Formatting Microsoft TV Content with Styles and Style Sheets

In This Chapter

- Where to Find Sample Content for This Chapter

- Microsoft TV's CSS Support

- Applying CSS Properties to Interactive TV Content

- Strategies for Implementing Styles: Inline Styles, Embedded Styles, and Linked Style Sheets

- Using DHTML to Dynamically Format Documents

With Microsoft TV, you can harness the power of cascading style sheets (CSS) properties to lay out and format interactive TV content. CSS is a simple style-sheet mechanism for the Web that enables content developers to position elements in a document, apply color to elements, and define fonts and font sizes for individual elements. For example, with CSS properties, you can position elements at an exact pixel location in a document without using cumbersome tables or transparent .gif images. You can also define specific font sizes, rather than using the traditional <font size=3> syntax. In addition, you can create a single style sheet to format multiple documents.

This chapter demonstrates several approaches for laying out interactive TV content with styles and style sheets, including using inline styles and linked style sheets. Perhaps equally important, it provides strategies for working around some of the limitations of Microsoft TV's CSS support.

WHERE TO FIND SAMPLE CONTENT FOR THIS CHAPTER

This chapter uses the Tailspin Toys sample content to demonstrate how to apply styles and style sheets to interactive TV content. We recommend that you view the sample files on the *Building Interactive Entertainment and E-Commerce Content for Microsoft TV* companion CD and Web site.

- To view the Tailspin Toys content on a set-top box running Microsoft TV or on the Web, go to *http://www.microsoft.com/tv/itvsamples*.

- To view the sample files associated with this chapter, see the "Formatting Microsoft TV Content with Styles and Style Sheets" topic in the "Microsoft TV Design Guide" section of the companion CD.

- For HTML source files of the toys template, see the "Templates" section of the companion CD.

- For more information about the dynamic HTML (DHTML) properties supported by Microsoft TV, see the "Microsoft TV Programmer's Guide" on the companion CD.

MICROSOFT TV'S CSS SUPPORT

Microsoft TV's CSS support offers great flexibility in formatting and laying out interactive TV content. However, designers should bear in mind that Microsoft TV does not support all the CSS1 properties defined by the World Wide Web Consortium's CSS1 recommendation and only a limited subset of CSS2 properties defined by the World

Wide Web Consortium's CSS2 recommendation. In addition, the *style* and *class* properties, which are integral to the use of CSS1, are supported by only a small subset of HTML elements.

■ To find out which CSS1 and CSS2 properties are supported by Microsoft TV, see the "CSS1 for Microsoft TV" topic in the "Microsoft TV Programmer's Guide" section of the companion CD.

■ In general, the style and class properties can be applied reliably only to [...] HTML elements and [...] Microsoft TV" topic [...] f the companion CD.

APPLY[...] [...]TERACTIVE TV C[...]

Usin[...] explain how to apply CSS1 and [...] in Toys content is designed to e[...] ling viewers to purchase an adve[...] ng a commercial for Swingin' Gur[...] /hen a viewer clicks the Go Inte[...] s shown in Figure 8.1.

Figure 8.1 *The interactive TV overlay for Tailspin Toys.*

When the viewer selects the Order Now button, a new page appears on the screen that lets the viewer purchase the doll online, as shown in Figure 8.2.

Figure 8.2 *The first purchase page for Swingin' Gurby.*

STRATEGIES FOR IMPLEMENTING STYLES: INLINE STYLES, EMBEDDED STYLES, AND LINKED STYLE SHEETS

With Microsoft TV, you can apply styles to interactive TV content in several ways, including:

- Using inline styles

- Using embedded styles

- Using linked style sheets

This section covers each of these methods. It also discusses general strategies to follow when implementing styles for Microsoft TV content.

Using the Style Property with <div> and <font> Tags

We recommend that you only apply CSS properties within <div> and <font> tags because Microsoft TV supports the style and class properties for these HTML tags. As mentioned earlier in this chapter, Microsoft TV does not support the style and class properties for all HTML elements. However, this should not prevent you from accomplishing the page layout tasks you wish to accomplish. For example, the <table> tag does not support absolute positioning using the style or class properties, but you can enclose a <table> tag within <div> tags and then apply the style property to the division. The following code snippet shows how this is done:

```
<html>
<head>
<title>CSS Test</title>
</head>
<body>
<div style="position:absolute;top:100;left:100;">
<table width="200" border="1">
<tr>
<td>CSS Test</td>
</tr>
</table>
</div>
</body>
</html>
```

> **NOTE** Apply CSS properties only to <div> and <font> tags.

Using Inline Styles for Absolute Positioning

Using absolute positioning is advantageous for Microsoft TV because the Microsoft TV window, unlike a computer-based window, cannot be resized. Microsoft TV always displays content in a fixed window size of 560 pixels by 420 pixels in TV mode. Another advantage is speed; using absolute positioning to lay out a page is much faster than using tables. Finally, HTML code created with absolute positioning is much easier to read and maintain than code created with embedded table structures.

One of the best ways to implement absolute positioning is with inline styles. To create an inline style, add the style property to a specific instance of an HTML element, as shown in the following syntax:

```
<tag style="attribute:value;attribute:value;...."></tag>
```

The following code, shown in bold, demonstrates how an inline style is used to apply absolute positioning for the toys_feature1.html page of the Tailspin Toys content. In this case, the Swingin' Gurby text is positioned 20 pixels from the top of the document and 14 pixels from the left of the document.

```
<body background="images/tile.gif" leftmargin="0" topmargin="0">

<div style="position:absolute;top:20;left:14">
<font class="clsTitle"><b>SWINGIN' GURBY</b></font>  .
</div>
```

Figure 8.3 shows how the Swingin' Gurby text appears on the page.

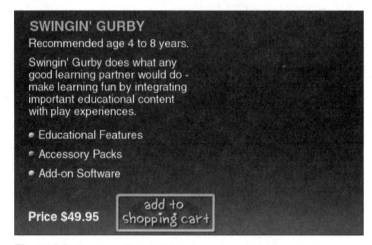

Figure 8.3 *The toys_feature1.html page.*

The following code shows how absolute positioning is used throughout the toys_feature1.html page to position HTML elements in the document:

```
<html>
<head>

<link rel="stylesheet" type="text/css" href="cssTV.css">

<script type="text/JavaScript">
var bName = navigator.appName
if (bName == "Microsoft Internet Explorer")
{
    document.write("<link rel='stylesheet' type='text/css'
        href='cssPC.css'>");
}
</script>

<title>Swingin' Gurby</title>
</head>
```

```
<body background="images/tile.gif" leftmargin="0" topmargin="0">

<div style="position:absolute;top:20;left:14;">
<font class="clsTitle"><b>SWINGIN' GURBY</b></font>
</div>

<div style="position:absolute;top:50;left:14;width:270">
<font class="clsDescription">Recommended age 4 to 8 years.</font>
</div>

<div style="position:absolute;top:80;left:14;width:270">
<font class="clsDescription">Swingin' Gurby does what any good learning
partner would do - make learning fun by integrating important educational
content with play experiences.</font>
</div>

<div style="position:absolute;top:190;left:14;width:260">
<a href="toys_sample1.html" target="content" nextright="fullbtn">
<img src="images/gbutton.gif" width=10 height=10 alt="" border="0">
  Educational Features</a>
</div>

<div style="position:absolute;top:220;left:14;width:260">
<a href="toys_sample1.html" target="content" nextright="fullbtn">
<img src="images/pbutton.gif" width=10 height=10 alt="" border="0">
  Accessory Packs</a>
</div>

<div style="position:absolute;top:250;left:14;width:260">
<a href="toys_sample1.html" target="content" nextright="fullbtn">
<img src="images/ybutton.gif" width=10 height=10 alt="" border="0">
  Add-on Software</a>
</div>

<div style="position:absolute;top:310;left:14;">
<font class="clsPrice"><b>Price $49.95</b></font>
</div>

<div style="position:absolute;top:296;left:144;width:260">
<a href="toys_sample1.html" target="content" selected>
<img src="images/cart.gif" width=148 height=48 alt="" border="0"></a>
</div>

</body>
</html>
```

In many cases, you can reduce the amount of code on a page by using embedded styles or a linked style sheet, as described in the following sections.

Using Embedded Style Sheets

To use an embedded style sheet, you define a style block, which is specified by the <style> tag, usually in the <head> section of the document. This block consists of a set of style rules, in which each rule defines a style for an HTML element or group of elements. A style rule has two parts:

- A selector that identifies an HTML element or group of elements

- A declaration of the style attributes to be applied to that selector

The generic syntax for a style rule is as follows:

```
selector {attribute:value;attribute:value;....}
```

To create an embedded style sheet, embed the styles between <style> tags as shown in the following code sample:

```
<html>
<head>
<style>
<!--
A {font-family:Arial,Verdana,Helvetica;font-size:13pt;color:#E0E0E0}
.clsTitle {font-family:Arial,Verdana,Helvetica;
font-size:16pt;color:#2BDB39}
.clsDescription {font-family:Arial,Verdana,Helvetica;
font-size:13pt;color:#E0E0E0}
.clsPrice {font-family:Arial,Verdana,Helvetica;
font-size:14pt;color:#E0E0E0}
-->
</style>
<title>Embedded Style Sample</title>
</head>

<body>
```

When defining styles for HTML elements such as <h1> or <p>, you simply specify the element, followed by the cascading style sheets (CSS) attributes within the "{ }" brackets as below:

```
H1 {font-family:Verdana, Helvetica, Sans Serif;color:blue}
```

To define a style independent of an HTML element, you create a class, followed by the attributes within the "{ }" brackets. When you create a class independent of an element, you can apply the class to format any text in the document. The following code illustrates several classes created for the toys_feature1.html document:

```
<style>
<!--
A {font-family:Arial,Verdana,Helvetica;font-size:13pt;color:#E0E0E0}
.clsTitle {font-family:Arial,Verdana,Helvetica;
font-size:16pt;color:#2BDB39}
.clsDescription {font-family:Arial,Verdana,Helvetica;
font-size:13pt;color:#E0E0E0}
.clsPrice {font-family:Arial,Verdana,Helvetica;
font-size:14pt;color:#E0E0E0}
-->
</style>
```

The classes are applied to text in the document, as shown in bold in the following code:

```
<body background="images/tile.gif" leftmargn="0" topmargin="0">

<div style="position:absolute;top:20;left:14;">
<font class="clsTitle"><b>SWINGIN' GURBY</b></font>
</div>

<div style="position:absolute;top:50;left:14;width:270">
<font class="clsDescription">Recommended age 4 to 8 years.</font>
</div>

<div style="position:absolute;top:80;left:14;width:270">
<font class="clsDescription">Swingin' Gurby does what any good learning
partner would do - make learning fun by integrating important educational
content with play experiences.</font>
</div>

<div style="position:absolute;top:190;left:14;width:260">
<a href="toys_sample1.html" target="content" nextright="fullbtn">
<img src="images/gbutton.gif" width=10 height=10 alt="" border="0">
  Educational Features</a>
</div>

<div style="position:absolute;top:220;left:14;width:260">
<a href="toys_sample1.html" target="content" nextright="fullbtn">
<img src="images/pbutton.gif" width=10 height=10 alt="" border="0">
  Accessory Packs</a>
</div>

<div style="position:absolute;top:250;left:14;width:260">
<a href="toys_sample1.html" target="content" nextright="fullbtn">
<img src="images/ybutton.gif" width=10 height=10 alt="" border="0">
  Add-on Software</a>
</div>
```

(continued)

```
<div style="position:absolute;top:310;left:14;">
<font class="clsPrice"><b>Price $49.95</b></font>
</div>

<div style="position:absolute;top:296;left:144;width:260">
<a href="toys_sample1.html" target="content">
<img src="images/cart.gif" width=148 height=48 alt="" border="0"></a>
</div>

</body>
</html>
```

Using Linked Style Sheets

Embedded style sheets work wonderfully, but they are limited in scope to the current document. A more global way to apply styles is to create a CSS and then link this style sheet to multiple documents. In fact, that's the way the toys_feature1.html document is formatted. The <link> tag will link a style sheet to a Web page as shown in the following code snippet:

```
<link rel="stylesheet" type="text/css" href="stylesheetName">
```

A linked style sheet consists of a set of style rules, exactly like an embedded style sheet. The following code sample shows the cssTV.css style sheet created for the Tailspin Toys content:

```
A {font-family:Arial,Verdana,Helvetica;font-size:18pt;color:#E0E0E0}

.clsTitle {font-family:Arial,Verdana,Helvetica;
font-size:22pt;color:#2BDB39}

.clsDescription {font-family:Arial,Verdana,Helvetica;
font-size:18pt;color:#E0E0E0}

.clsPrice {font-family:Arial,Verdana,Helvetica;
font-size:20pt;color:#E0E0E0}
```

To create a style sheet, you generate a document using an HTML editor, Notepad, or a word processor. Unlike an HTML document, the CSS document contains only styles, much like the sample code above. After you have created the necessary styles, you save the file with a .css extension. For example, the style sheet used to format toys_feature1.html is called cssTV.css. The text in bold in the following code sample illustrates how cssTV.css is linked to the toys_feature1.html document. Classes defined in a linked style sheet are applied in the same way that classes are applied for an embedded style sheet. The applied classes are also shown in bold in the following code sample:

```
<html>
<head>
<link rel="stylesheet" type="text/css" href="cssTV.css">
<title>SWINGIN' GURBY</title>
</head>

<body background="images/tile.gif" leftmargin="0" topmargin="0">

<div style="position:absolute;top:20;left:14;">
<font class="clsTitle"><b>SWINGIN' GURBY</b></font>
</div>

<div style="position:absolute;top:50;left:14;width:270">
<font class="clsDescription">Recommended age 4 to 8 years.</font>
</div>

<div style="position:absolute;top:80;left:14;width:270">
<font class="clsDescription">Swingin' Gurby does what any good learning
partner would do - make learning fun by integrating important educational
content with play experiences.</font>
</div>
```

Dynamically Applying Style Sheets

One advantage of linked style sheets is that you can use JavaScript to dynamically apply a style sheet based on a given set of conditions. For example, a trick used for sample content in this book is to apply the cssPC.css style sheet for designing on the computer and apply the cssTV.css style sheet for displaying content on TV. The cssPC.css style sheet, in effect, approximates on the computer at design time what you'll see when the content is rendered by Microsoft TV. The following code shows the cssPC.css style sheet created for Tailspin Toys:

```
A {font-family:Arial,Verdana,Helvetica;font-size:13pt;color:#E0E0E0;
font-weight:bold;text-decoration:none}
.clsTitle {font-family:Arial,Verdana,Helvetica;
font-size:16pt;color:#2BDB39}
.clsDescription {font-family:Arial,Verdana,Helvetica;
font-size:13pt;color:#E0E0E0;font-weight:bold}
.clsPrice {font-family:Arial,Verdana,Helvetica;
font-size:14pt;color:#E0E0E0}
```

As shown in bold in the following code sample, JavaScript is added to the toys_feature1.html document to attach the cssPC.css style sheet when the document

is loaded. If the Microsoft Internet Explorer appName property is not detected, however, the cssPC.css style sheet is not loaded and the cssTV.css style sheet is used.

```
<html>
<head>

<link rel="stylesheet" type="text/css" href="cssTV.css">

<script type="text/JavaScript">
var bName = navigator.appName
if (bName == "Microsoft Internet Explorer")
{
    document.write("<link rel='stylesheet' type='text/css'
       href='cssPC.css'>");
}
</script>
```

If the dual style sheet solution seems confusing right now, don't worry. This
solution for dynamically adjusting font sizes for the computer and the TV is covered
in more detail in Chapter 10, "Creating Text for Microsoft TV."

USING DHTML TO DYNAMICALLY FORMAT DOCUMENTS

Microsoft TV provides a subset of DHTML properties that you can use to dynamically
format documents. For example, with Microsoft TV's subset of DHTML, you can
dynamically move elements on a page to create animation. You can also hide or show
elements based on a user interaction. The following code snippet illustrates the
JavaScript that hides and shows the words "sample text" when the Hide or Show link
is clicked:

```
<html>
<head>
<script type="text/JavaScript">
function hide_show()
{
    if (document.all.sample_text.style.visibility == "visible")
        document.all.sample_text.style.visibility = "hidden";
    else
        document.all.sample_text.style.visibility = "visible";
}
</script>
<style>
.sample {visibility:hidden;color:red}
</style>
<title>Untitled</title>
```

```
</head>
<body>
<a href="JavaScript:void(0)" onClick="hide_show()">Hide or Show</a>
<div class="sample" id="sample_text">
Sample Text
</div>
</body>
</html>
```

Keep in mind that Microsoft TV's support for the properties of the style object is limited, and that the document.all.id.style.visibility statement is not supported by the specification defined by the Advanced Television Enhancement Forum (ATVEF). As a result, DHTML code using the document.all.id.style.visibility syntax may not work on all TV receivers built to ATVEF standards.

WHAT'S NEXT

This chapter demonstrated how to apply CSS1 and CSS2 properties to Microsoft TV content by using inline styles, embedded styles, and linked style sheets. As demonstrated in this chapter, you can use CSS properties to position elements on a Web page. In Chapter 9, "Selecting Colors for Microsoft TV Content," and Chapter 10, we will demonstrate how you can use CSS properties to define colors, fonts, and font sizes for Microsoft TV content.

Selecting Colors for Microsoft TV Content

In This Chapter

■   Where to Find Sample Content for This Chapter

■   Selecting and Adjusting Colors

■   Specifying Colors with HTML and Style Sheets

■   Testing Colors on Microsoft TV

Selecting colors for interactive TV content is challenging, because the colors you choose when designing content on a personal computer look different when shown on TV. Most colors are brighter on TV than on a personal computer. For example, bright red, which is perfectly satisfactory on a computer, tends to glow or bleed into other colors when shown on TV. Bright yellow on the TV creates a "crawling" effect. Pure white text tends to blur, while a pure white page background can actually cause the TV screen to bow on the sides.

Anyone who designs interactive TV content should be aware of these hazards. Designers also need to select colors that are considered suitable for TV by the National Television Standards Committee (NTSC). These colors are called NTSC-safe colors.

This chapter demonstrates a proven method for designing pages with NTSC-safe colors. It also discusses techniques used by graphic artists to adjust NTSC-safe colors to fit within the range of their designs. We expect that most readers of this chapter will be graphic artists who are familiar with a high-end graphics program, such as Adobe Photoshop.

WHERE TO FIND SAMPLE CONTENT FOR THIS CHAPTER

Using the Tailspin Toys sample content introduced in Chapter 8, "Formatting Microsoft TV Content with Styles and Style Sheets," we will demonstrate how to select NTSC-safe colors for interactive TV content. The sample files on the companion CD and Web site will help clarify this process.

- To view the Tailspin Toys content on a set-top box running Microsoft TV or on the Web, go to *http://www.microsoft.com/tv/itvsamples*.

- To view the sample files associated with this chapter, see the "Selecting Colors for Microsoft TV Content" topic in the "Microsoft TV Design Guide" section of the companion CD.

- For HTML source files of the toys template, see the "Templates" section of the companion CD.

SELECTING AND ADJUSTING COLORS

Before we jump into a discussion of how to apply colors to Microsoft TV content, it helps to have some background information about color differences between the computer and the TV.

About NTSC, True Color, High Color, and Browser-Safe Colors

Microsoft TV in North America supports only NTSC-safe colors. The computer, however, supports color standards such as High Color and True Color. The High Color palette, for example, enables you to select from up to 65,536 colors, while the True Color palette enables you to select from up to 16,777,216 colors for 24-bit graphics and 4,294,967,296 colors for 32-bit graphics. The "NTSC color space," as it's called, also offers a vast assortment of colors, but this assortment varies from those colors offered by the True Color and High Color standards.

An NTSC filter or converter can be used to reconcile the differences between High Color and True Color on the computer, as well as the NTSC color space on TV. Here is how the process works: When you design interactive TV content, you design it on a computer that has display settings configured for True Color or High Color. Normally, the initial design is a mock-up created in a graphics program such as Photoshop. After you create the mock-up, you apply an NTSC filter or converter to the mock-up to convert the colors to NTSC-safe colors for Microsoft TV.

Should You Worry About Browser-Safe Colors for Microsoft TV?

Most likely, if you have designed Web content, you are familiar with the concept of browser-safe colors. Browser-safe colors are the 216 colors that are common to the Microsoft Windows, Windows 95, Windows 98, Windows NT, Windows 2000, and Macintosh operating systems. These colors are known not to dither, regardless of the operating system and browser on which they are displayed. Browser-safe colors are a design consideration for Microsoft TV only if you plan to display content on both the computer and the TV.

This book focuses on creating interactive TV content specifically for Microsoft TV. Accordingly, its sample content uses colors that are optimized for TV without regard to browser-safe color constraints. As a general rule, we do not recommend using browser-safe colors for interactive TV designs. After all, TV content would look mighty dull if limited to 216 colors.

Setting the Computer for True Color or High Color

When designing interactive TV content, it is advisable to set your computer display settings for True Color or High Color. To do this, select the True Color or High Color palette in the Display settings of the Windows Control Panel.

To select the True Color or High Color palette on the computer:

1. Click the Start button on the taskbar, point to Settings, and then click Control Panel.

2. Double-click the Display icon, and then click the Settings tab.

3. In the Color Palette dialog box, select either True Color or High Color.

Using Graphics Software with an NTSC Filter or Converter

To design interactive TV content, use a graphics program such as Adobe Photoshop that includes an NTSC filter or converter that maps colors on the computer to NTSC-safe colors. The NTSC filter generally reduces the brightness or saturation of certain colors to make them suitable for display on TV.

Using a Computer-to-Video Scan Converter and an NTSC Monitor

A good way to design interactive TV content is to actually view it on the TV while you are creating it. To do this, you can use a computer-to-video scan converter to connect a computer to a TV monitor. For help in selecting a computer-to-video scan converter, see the "Third Party Vendors" topic in the "Resources" section of the companion CD.

Creating a Design Mock-up

It is nearly always useful to create mock-ups of the pages you intend to include in your interactive TV content. A mock-up is a visual rendering of the proposed design of the page. Design mock-ups are usually created by graphic artists, who are typically in charge of designing interactive content. Figure 9.1 shows a mock-up of the Tailspin Toys content that was created using Adobe Photoshop. (Note that an image has been added to the TV control area for effect.)

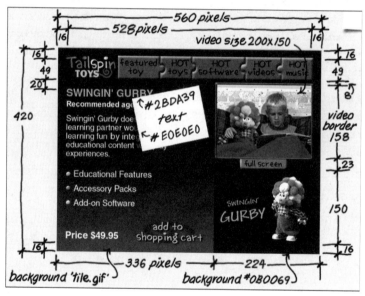

Figure 9.1 *A mock-up page created with a graphics program.*

Applying the NTSC Filter or Converter

After a mock-up has been created, you can apply the NTSC filter or converter to ensure that your colors are NTSC-safe. Always apply the filter or converter to all elements of a mock-up. For example, with Photoshop, apply the filter to all layers on the page. Instructions for applying the filter or converter will vary among software programs.

Adjusting Colors Applied by the NTSC Filter or Converter

The NTSC filter or converter ensures that colors are NTSC-safe and thus suitable for TV. Nevertheless, in some cases the colors selected by the filter or converter may not be to your liking. In addition, pure white is not affected by the filter. For these reasons, after applying an NTSC filter or converter, it is often necessary to readjust colors in the mock-up. Figure 9.2 shows the three categories of colors involved in the Tailspin Toys image: the top row depicts the original Red, Green, Blue (RGB) values; the middle row shows the colors adjusted by the NTSC filter; and the bottom row shows the color values readjusted by the graphic artist.

255/0/0	250/0/255	0/233/204	255/200/0	0/255/0	brights
181/0/0	210/0/210	0/189/166	193/151/2	0/210/0	NTSC filter applied
195/162/41	231/51/199		250/156/17	0/177/37	color adjustment

Figure 9.2 *Original colors (top), NTSC-safe colors (center), and adjusted NTSC-safe colors (bottom).*

Here is how the graphic artist arrived at the final colors for the toolbar of Tailspin Toys:

- **Red**—The initial pure value of R=255; G=0; B=0 was darkened and desaturated by the NTSC filter. The resulting color was too dark for use in the design. In order to have a lighter safe value, the hue, saturation, and lightness were slightly adjusted. When the NTSC filter was again applied, the new red value had no change—resulting in a safe color.

- **Magenta**—The initial pure value of R=255; G=0; B=255 was darkened and desaturated by the NTSC filter. The resulting color appeared too dark when viewed. To lighten the color, the hue, saturation, and lightness were adjusted.

- **Aqua**—The starting value of R=0; G=233; B=204 was darkened and desaturated by the NTSC filter. The resulting color was acceptable for use in the design.

- **Yellow**—The original value of R=255; G=200; B=0 was darkened and desaturated by the NTSC filter. The resulting color was too dark and muddy for the design. In order to have a lighter safe value, the hue, saturation, and lightness were slightly adjusted. When the NTSC filter was again applied, the new yellow value had no change—resulting in a safe color.

■ **Green**—The initial pure value of R=0; G=255; B=0 was darkened and desaturated by the NTSC filter. The resulting color still appeared too bright when viewed on the TV. To tone down the color, the hue, saturation, and lightness were slightly adjusted.

■ **White**—The pure value of R=255; G=255; B=255 is forgotten by the NTSC filter. Make sure to adjust white by 10 to 20 percent.

The NTSC Filter and Pure White

The NTSC filter in Photoshop does not affect pure white (RGB value R=255; G=255; B=255). Nevertheless, usability studies show that TV viewers dislike viewing content on a pure white background. In addition, pure white text tends to blur on TV, and a pure white background can cause distortion. When designing for interactive TV, you should manually change pure white backgrounds or text colors from pure white (hexadecimal value #FFFFFF) to a more subdued, NTSC-safe color, such as #FEFEFE, #E0E0E0, or #CCCCCC. The color you select to replace pure white will depend on your design.

The Zen of Color Selection

Unfortunately, there is no formula for making color adjustments to colors imposed by an NTSC filter or converter. How a color appears on TV depends on the amount of color, the neighboring colors, and the background color. For example, a big block of red can vibrate as much as a thin line of the same color. For best results, always view interactive content on a TV monitor.

As a general rule, colors that tend to work best for interactive TV designs are the cool tones of blue, green, and violet. The warm tones of red, yellow, and orange tend to need more adjustment to avoid distortion such as bleeding or crawling.

When adjusting colors for TV, some colors may need only the hue adjusted, while others may need adjustments only for saturation or brightness. Some colors may need adjustments for all three of these values. For example, when the NTSC filter in Photoshop is applied to a bright yellow (R=255; G=200; B=0), it decreases the brightness by about 30 percent, dropping the yellow to the R=193; G=151; B=0 RGB value, which can appear muddy or a drab olive green on TV. To create an acceptable yellow for TV with Photoshop, you can adjust the hue (-11), saturation (-19), and lightness (+13) to create an RGB value of R=220; G=136; B=13. When the NTSC filter is reapplied, there is no change in the color. After readjusting colors, it is necessary to reapply the NTSC filter to ensure the adjusted colors are within the NTSC range. It may be necessary to reiterate the "color adjustment/apply NTSC filter" process to get the results you want.

SPECIFYING COLORS WITH HTML AND STYLE SHEETS

After the colors in the design mock-up are finalized, you can use the eyedropper in your graphics program to determine the RGB color values for the design elements that you implement with HTML code. Next, you convert the RGB values to hexadecimal values and apply the hexadecimal values to the HTML elements. The following sections describe how this is done.

Using the Eyedropper to Determine RGB Values for HTML Elements

After adjusting colors, reapplying the NTSC filter, and finalizing mock-ups, you can use the eyedropper in your graphics program to determine RGB color values for elements of the design that will be implemented with HTML code. (Most high-end graphics programs contain an eyedropper tool that you can position over a graphic element to determine the RGB value of its color.)

For the Tailspin Toys content, for example, the text colors are implemented using HTML. Therefore, to determine the RGB value of the text on the mock-up page, you place the eyedropper over the text and jot down the indicated RGB values. Note that instructions for the eyedropper vary with each graphics program.

Converting Values to Hexadecimal

After determining the RGB values of the elements on the mock-up that will be implemented by HTML, the next step is to convert the RGB values to hexadecimal values. Hexadecimal values are the standard way of specifying colors for Web pages. For example, the RGB value of R=255; G=0; B=0 (pure red) equates to a hexadecimal value of #FF0000. There are a variety of ways to convert RGB values to hexadecimal values. Here is an easy method that uses the Windows calculator.

To convert values from RGB to hexadecimal using the Windows Calculator:

1. Click Start on the Windows taskbar, point to Programs, and then point to Accessories.

2. Click Calculator.

3. From the View menu of the Calculator window, click Scientific.

4. In the box at the top right of the window, type the first number of the RGB value. For example, for pure red, type 255.

5. Click the Hex radio button to show the hexadecimal value.

6. Repeat for each RGB value. For example, for pure red (RGB value R=255; G=0; B=0), the hexadecimal value is #FF0000. Note that the hexadecimal value is always six characters.

Specifying Colors for the HTML Elements of Tailspin Toys

Once you have converted RGB values to hexadecimal, you can specify hexadecimal values for the HTML elements on your page. Figure 9.3 shows how a graphic artist can specify hexadecimal values for the Web developer. As shown in the illustration, hexadecimal values are set for the text on the page. Note that the colors for the graphics elements, including the tiled background, the bullets, and the "add to shopping cart" image, have already been defined in the graphics program.

Figure 9.3 *The toys_feature1.html page of Tailspin Toys.*

Color for the text on the toys_feature1.html page is defined through the use of a TV style sheet, called cssTV.css, as shown in the following code sample. Notice that the color attribute for the clsTitle class is defined as #2BDA39, which is a shade of green, while the color attribute for the <a> tag and the clsDescription and clsPrice classes is defined as #E0E0E0, which is a muted shade of white.

```
A {font-family:Arial,Verdana,Helvetica;font-size:18pt;color:#E0E0E0}
.clsTitle {font-family:Arial,Verdana,Helvetica;
font-size:22pt;color:#2BDA39}

.clsDescription {font-family:Arial,Verdana,Helvetica;
font-size:18pt;color:#E0E0E0}

.clsPrice {font-family:Arial,Verdana,Helvetica;
font-size:20pt;color:#E0E0E0}
```

Linking the Style Sheet and Applying the Classes

Following is the complete code for the toys_feature1.html page of Tailspin Toys. Notice that the cssTV.css style sheet is linked at the beginning of the document using the <link rel="stylesheet" type="text/css" href="cssTV.css"> statement. Also notice that each font tag in the document is assigned a class defined in the cssTV.css style sheet.

```
<html>
<head>

<link rel="stylesheet" type="text/css" href="cssTV.css">

<script language="JavaScript">
var bName = navigator.appName
if (bName == "Microsoft Internet Explorer")
{
    document.write("<link rel='stylesheet' type='text/css'
        href='cssPC.css'>");
}
</script>

<title>Swingin' Gurby</title>
</head>

<body background="images/tile.gif" leftmargin="0" topmargin="0">

<div style="position:absolute;top:20;left:14;">
<font class="clsTitle"><b>SWINGIN' GURBY </b></font>
</div>

<div style="position:absolute;top:50;left:14;width:270">
<font class="clsDescription">Recommended age 4 to 8 years.</font>
</div>

<div style="position:absolute;top:80;left:14;width:270">
<font class="clsDescription">
Swingin' Gurby does what any good learning partner would do - make
learning fun by integrating important educational content with play
experiences.
</font>
</div>

<div style="position:absolute;top:190;left:14;width:260">
<a href="toys_sample1.html" target="content" nextright="fullbtn">
<img src="images/gbutton.gif" width=10 height=10 alt="" border="0">
  Educational Features</a>
</div>

<div style="position:absolute;top:220;left:14;width:260">
<a href="toys_sample1.html" target="content" nextright="fullbtn">
<img src="images/pbutton.gif" width=10 height=10 alt="" border="0">
  Accessory Packs</a>
</div>
```

(continued)

```
<div style="position:absolute;top:250;left:14;width:260">
<a href="toys_sample1.html" target="content" nextright="fullbtn">
<img src="images/ybutton.gif" width=10 height=10 alt="" border="0">
  Add-on Software</a>
</div>

<div style="position:absolute;top:310;left:14;">
<font class="clsPrice"><b>Price: $49.95</b></font>
</div>

<div style="position:absolute;top:296;left:144;width:260">
<a href="toys_sample1.html" target="content" selected>
<img src="images/cart.gif" width=148 height=48 alt="" border="0"></a>
</div>

</body>
</html>
```

> **NOTE** In the preceding code, JavaScript is used to dynamically apply the cssPC.css style sheet for Microsoft Internet Explorer. This solution is covered in more detail in the next chapter.

TESTING COLORS ON MICROSOFT TV

The only true test of colors is to view them on TV. Therefore, after the pages for your interactive TV content have been designed and coded, it is a good idea to post them to an external Web server and view them with a Microsoft TV–based receiver such as the WebTV Plus Receiver. Following are instructions for viewing content on the WebTV Service with a WebTV Plus Receiver.

To view content with a WebTV Plus Receiver:

1. Turn on your WebTV Plus Receiver.

2. Click View on the keyboard to connect to the WebTV Service in Web mode.

3. Click Options on the keyboard, and select the Go To button on the screen.

4. In the address box, type the URL where you posted the content.

5. Press the Return key on the keyboard.

WHAT'S NEXT

This chapter demonstrated how to select NTSC-safe colors for Microsoft TV content. It also described how to specify colors for interactive TV content using style sheets. In the next chapter, we will explain how to use styles and style sheets to control fonts and font sizes for Microsoft TV content.

Chapter 10

Creating Text for Microsoft TV

In This Chapter

■   Where to Find Sample Content for This Chapter

■   What to Avoid When Creating Text for Microsoft TV

■   How the Microsoft TV Proxy Server Handles Fonts

■   How to Use Styles to Control Font Sizes

■   How to Reconcile Font Sizes on the Computer and the TV

To create TV-friendly content, it is important to use fonts and font sizes that are optimal for TV viewing. Research shows that TV viewers in North America, on average, watch television from a distance of nine feet, so it is important to create text that is large enough to be read from across the room. Choosing the correct fonts is also critical, since studies have shown that serif fonts, such as Times New Roman, are more difficult to read on TV than sans serif fonts, such as Helvetica.

This chapter demonstrates how to create TV-friendly text for Microsoft TV content by covering the following topics:

- How to avoid pitfalls in selecting fonts and font sizes for Microsoft TV

- How Microsoft TV handles fonts

- How to use font families supported by Microsoft TV

- How to use style sheets to format text for Microsoft TV content

- How to use style sheets to reconcile font sizes between the computer and the TV

WHERE TO FIND SAMPLE CONTENT FOR THIS CHAPTER

This chapter uses the Exotic Excursions sample content to demonstrate how to apply styles and style sheets to interactive TV content. To get the most out of the material in this chapter, we recommend that you view the sample files on the *Building Interactive Entertainment and E-Commerce Content for Microsoft TV* companion CD and Web site.

- To view the Exotic Excursions content on a set-top box running Microsoft TV or to view the content on the Web using your computer, go to *http://www.microsoft.com/tv/itvsamples*.

- To view the sample files associated with this chapter, see the "Creating Text for Microsoft TV" topic in the "Microsoft TV Design Guide" section of the companion CD.

- To access source files of the Exotic Excursions content, see the "Templates" section of the companion CD.

An Overview of the Exotic Excursions Content

This chapter uses the Exotic Excursions content as an example of how to create text for Microsoft TV using styles and style sheets. The Exotic Excursions content is designed to extend the value of a travel show. During the show, interactive TV links appear on the TV screen. These links enable the TV viewer to click the Go Interactive button and view interactive TV content. As shown in Figure 10.1, the Exotic Excursions interactive content offers packaged tour dates for geographical areas featured in the show. It also lets the viewer make hotel reservations, view slide shows of the areas, review travel tips for the areas, and shop for budget travel tours.

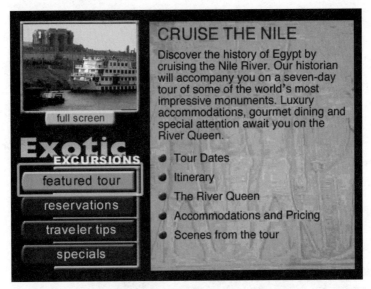

Figure 10.1 *The Exotic Excursions interactive TV content.*

WHAT TO AVOID WHEN CREATING TEXT FOR MICROSOFT TV

When creating text for Microsoft TV, there are a few common pitfalls you should avoid, as described in the following sections.

Small Text and Serif Fonts are Difficult to Read

Usability studies have shown that text smaller than 18 points is difficult for TV viewers to read. As a result, the Microsoft TV proxy server automatically converts all HTML-based text to Helvetica or, in some cases, to a fixed-width font called Monaco that is similar to Courier. Unless a font size is fixed via a cascading style sheets (CSS) style, the Microsoft TV proxy server automatically enlarges fonts smaller than 18 points to 18 points, and enlarges other fonts on the page accordingly. However, the Microsoft TV proxy server does not reformat text embedded in graphics, so avoid embedding text smaller than 16 points in images. For HTML-based text, you should always use text that is 16 points or larger.

Usability studies also show that serif fonts (the ones with the short lines attached to the strokes of the letters, such as Times New Roman) are more difficult to read on TV than sans serif fonts (the ones without the short lines attached to the letter strokes, such as Helvetica) because serif fonts tend to blur. As a result, when creating images that require text, use either Helvetica or a sans serif font that works well with Helvetica.

Design Breaks Down When Users Change Font Size

When creating text for Microsoft TV content, you should be aware that Microsoft TV allows users to change their font size settings to Small, Medium, or Large. Unless you design for all three font size settings or explicitly format font sizes using styles, the integrity of your page design can be ruined when the user changes the font size setting to Large. Also, keep in mind that the Microsoft TV client does not support page scrolling in TV mode. Thus, in some cases, the user can change the font size to Large and actually push text out of the visible boundary of the TV screen. For example, Figure 10.2 shows a sample page for Exotic Excursions with the font size set to Medium.

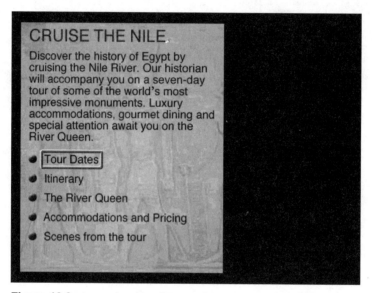

Figure 10.2 *A sample page of Exotic Excursions with the font size set to Medium.*

Figure 10.3 shows the same page with the font size set to Large. In this case, the enlarged text is pushed into other text on the page because it has been positioned with absolute positioning. If absolute positioning were not used for this page, the text could actually extend beyond the viewable area of the TV screen. And, because Microsoft TV does not support scrolling of Web pages in TV mode, viewers would have no way to view the text unless they changed their font size to Medium or Small.

Figure 10.3 *A sample page with the font size set to Large.*

To prevent TV viewers or Microsoft TV from disrupting the integrity of your design, the best solution is to fix fonts with styles. This solution is discussed under the heading "How to Use Styles to Control Font Sizes" later in this chapter.

Bright Text Can Cause Distortion

When formatting text for interactive TV content, avoid using bright colors that cause unwanted effects such as glowing, bleeding, or crawling. For more information about selecting colors for TV, see Chapter 9, "Selecting Colors for Microsoft TV Content."

HOW THE MICROSOFT TV PROXY SERVER HANDLES FONTS

Microsoft TV automatically converts fonts to Helvetica (and in some cases Monaco) and converts all fonts smaller than 18 points to 18 points, unless the font sizes are fixed with a CSS style. Font sizes larger than 18 points are scaled accordingly, with the exception of the "size=7" setting, which does not scale the font size. For example, Figure 10.4 illustrates a sample page as shown on the computer.

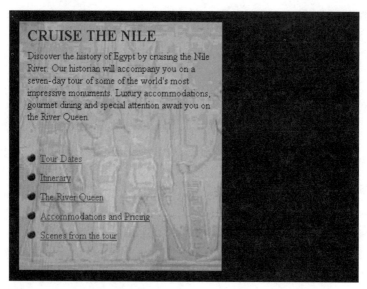

Figure 10.4 *A sample page as shown on the computer.*

Figure 10.5 shows the same page on Microsoft TV. In this case, the font is changed from Times New Roman to Helvetica, and the font sizes are enlarged for TV display.

Figure 10.5 *A sample page as shown on Microsoft TV.*

The following table demonstrates how Microsoft TV translates font sizes for the Small, Medium, and Large font settings. The top portion of the table represents the possible font sizes (<font size=1> - <font size=7>) that you can specify without

Cascading Style Sheets level 1 (CSS1) properties in HTML. For example, <font size=3> would result in a 20-point font being used if the TV viewer has the size setting set to Medium on the TV receiver. The left portion of the table represents the possible font size settings on the TV receiver.

	1	2	3	4	5	6	7
Small:	14	16	18	20	24	26	28
Medium:	16	18	20	22	26	30	34
Large:	18	20	22	24	28	34	42

HOW TO USE STYLES TO CONTROL FONT SIZES

When designing for the Microsoft TV client, you cannot change the font family because the Microsoft TV client supports only Helvetica and a fixed-width font called Monaco. But you can control font sizes through the use of styles. By controlling font sizes with styles, you can protect the integrity of your design so that neither the Microsoft TV proxy server nor TV viewers can change the font size you specify. You can protect your specified font sizes in this way because, when fonts are formatted using styles, the style attributes are not overridden by the Microsoft TV client or by the user changing the font size setting. For example, the exotic_feature1.html page, as shown in the Figure 10.6, looks the same whether the user sets the font size to Small, Medium, or Large.

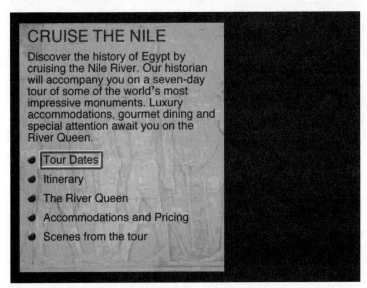

Figure 10.6 *Exotic_feature1.html as shown on Microsoft TV.*

The fonts for the exotic_feature1.html page of Exotic Excursions were formatted using the following cssTV.css style sheet. Notice that font size settings are specifically defined.

```
A {font-family:Arial,Verdana,Helvetica;font-size:18pt;color:#000000}

.clsTitle {font-family:Arial,Verdana,Helvetica;
font-size:24pt;color:#40194B}

.clsDescription {font-family:Arial,Verdana,Helvetica;
font-size:18pt;color:#000000}

.clsPrice {font-family:Arial,Verdana,Helvetica;
font-size:20pt;color:#E0E0E0}

.clsBack {font-family:Arial,Verdana,Helvetica;font-size:16pt;color:#68513f}
```

To attach the cssTV.css style sheet to the exotic_feature1.html page, the "<link rel="stylesheet" type="text/css" href="cssTV.css">" statement is used. To apply a style class, a class is specified within the <font> tag, as shown in bold in the following sample code:

```
<html>
<head>
<link rel="stylesheet" type="text/css" href="cssTV.css">
<script type="text/JavaScript">
var bname = navigator.appName
if (bname == "Microsoft Internet Explorer")
{
   document.write("<link rel='stylesheet'
      type='text/css' href='cssPC.css'>");
}
</script>

<title>Featured Tour Content</title>
</head>

<body bgcolor="black">

<div style="position:absolute;top:16;left:11;width:327;height:388;
background-image:url(images/back.jpg)">
</div>

<div style="position:absolute;top:25;left:24;">
<font class="clsTitle">CRUISE THE NILE</font>
</div>

<div style="position:absolute;top:60;left:24;width:305">
<font class="clsDescription">Discover the history of Egypt by cruising the
```

Nile River. Our historian will accompany you on a seven-day tour of some of
the world's most impressive monuments. Luxury accommodations, gourmet
dining and special attention await you on the River Queen.**</font>**
</div>

For more information about defining styles or attaching style sheets, see
Chapter 8, "Formatting Microsoft TV Content with Styles and Style Sheets."

HOW TO RECONCILE FONT SIZES ON THE COMPUTER AND THE TV

Using a style sheet to format text for TV is a great idea, but Figure 10.7 shows what
happens to the font sizes on the computer when you apply the cssTV.css style sheet.

Figure 10.7 *Exotic_feature1.html on a personal computer.*

As you can see, the fixed font sizes actually appear larger on the computer than
they do on Microsoft TV, which makes designing content on a computer somewhat
difficult. The following table shows the approximate relationship between the font
sizes on a personal computer and Microsoft TV.

Font Size on a Personal Computer	Equals Font Size on Microsoft TV
13 point Arial bold	18 point Helvetica
14 point Arial bold	20 point Helvetica
16 point Arial bold	22 point Helvetica

An easy way to reconcile font size differences between the computer and Microsoft TV is to create a style sheet that approximates on the computer what appears on the TV. For example, for the exotic_feature1.html page, we created a cssPC.css style sheet to handle font rendering on the computer. When this style sheet is attached to the exotic_feature1.html page, the page on the computer becomes very similar in appearance to the sample_style.html page on Microsoft TV. The following sample code shows the code for the cssPC.css style sheet:

```
A {font-family:Arial,Verdana,Helvetica;font-size:13pt;color:#000000;
font-weight:bold;text-decoration:none}

.clsTitle {font-family:Arial,Verdana,Helvetica;font-size:18pt;
font-weight:bold;color:#40194B;font-weight:bold}

.clsDescription {font-family:Arial,Verdana,Helvetica;
font-size:13pt;color:#000000;font-weight:bold}

.clsPrice {font-family:Arial,Verdana,Helvetica;
font-size:14pt;color:#E0E0E0;font-weight:bold}

.clsBack {font-family:Arial,Verdana,Helvetica;font-size:13pt;color:#68513f}
```

The following code sample shows how JavaScript is added to the exotic_feature1.html page to attach the cssPC.css style sheet:

```
<html>
<head>
<link rel="stylesheet" type="text/css" href="cssTV.css">
<script type="text/JavaScript">
var bName = navigator.appName
if (bName == "Microsoft Internet Explorer")
{
   document.write("<link rel='stylesheet' type='text/css'
      href='cssPC.css'>");
}
</script>
```

Figure 10.8 shows how the cssPC.css style sheet renders the exotic_feature1.html page on the computer. When the cssPC.css style sheet is attached, the fonts on the computer at design time closely resemble what you see on Microsoft TV.

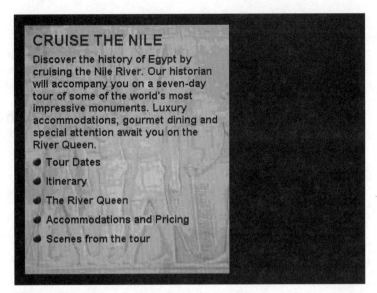

Figure 10.8 *Exotic_feature1.html on the computer with the cssPC.css style sheet attached.*

WHAT'S NEXT

This chapter demonstrated how to avoid common pitfalls when creating text for Microsoft TV content. It also demonstrated how to use styles and style sheets to avoid text conversion by the Microsoft TV proxy server and by the TV viewer. In the next chapter, we'll take a look at how to create and add images to Microsoft TV content.

Web content positioned over full-screen TV.

The TV object positioned over Web content.

The generic interactive TV template.

The generic template modified for "news on demand."

The Amy Jones Lifestyle content lets TV viewers see the recipe being prepared on the show.

During the Amy Jones Lifestyle show, TV viewers can request an e-mail copy of the recipe.

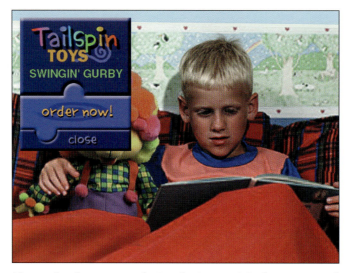

The overlay that appears during the Swingin' Gurby commercial.

Online shopping integrated with the Swingin' Gurby commercial.

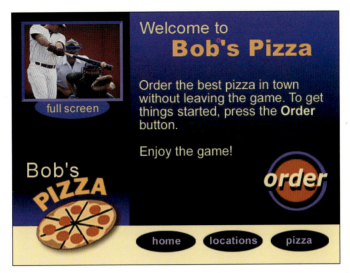

The Bob's Pizza application is integrated into a sporting event and lets TV viewers order a pizza online while watching a game.

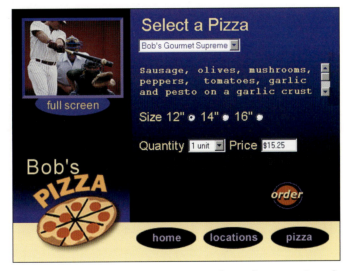

Using a remote control, TV viewers can choose from a variety of options when ordering a pizza.

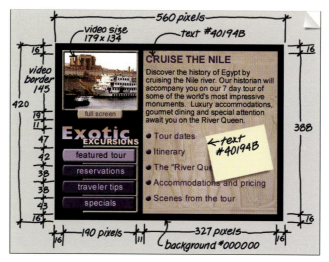

Before creating interactive TV content, the graphic artist creates a mockup and specifies the dimensions for page components.

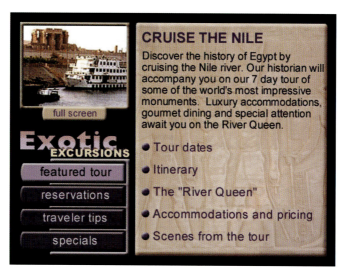

The Exotic Excursions content enhances a travel show, enabling TV viewers to make travel reservations online.

A form tied to a television commercial enables Humongous Insurance to collect leads from TV viewers.

With the Island Hopper News, TV viewers can get weather information anytime they want with a click of the remote. (This sample is not included with the book.)

With the Sports Updater, TV viewers can view the sports scores they want or create a personalized sports ticker. (This sample is not included with the book.)

With a presidential poll tied to a political debate, TV viewers can participate by voting online. (This sample is not included with the book.)

Chapter 11

Adding Images to Microsoft TV Content

In This Chapter

- Where to Find Sample Content for This Chapter
- Using Supported Image Types
- Adding Images to Microsoft TV Content
- Creating a User Interface with Images and JavaScript
- Adding Images as Background
- Using Image Maps

TV viewers are accustomed to viewing content that is designed for maximum visual appeal. As a result, graphic design should play a central role in any content designed for Microsoft TV. This chapter explains the image types supported by Microsoft TV, discusses the pitfalls of creating images for interactive TV, and shows how to create dynamic user interfaces by swapping images using JavaScript. As part of the discussion about swapping images, this chapter also shows how to preload images into the TV receiver's cache so that images load quickly in response to a user's selection.

WHERE TO FIND SAMPLE CONTENT FOR THIS CHAPTER

This chapter uses the Exotic Excursions sample content to demonstrate how to create images for Microsoft TV. To get the most benefit out of this chapter, it is recommended that you view the sample files on the *Building Interactive Entertainment and E-Commerce Content for Microsoft TV* companion CD and Web site.

- To view the Exotic Excursions content on a set-top box running Microsoft TV or to view the content on the Web using your computer, go to *http://www.microsoft.com/tv/itvsamples*.

- To view the sample files associated with this chapter, see the "Adding Images to Microsoft TV Content" topic in the "Microsoft TV Design Guide" section of the companion CD.

- For HTML source files of the Exotic Excursions template, see the "Templates" section of the companion CD.

An Overview of the Exotic Excursions Content

This chapter uses the Exotic Excursions content as an example of how to add images, preload images, and swap images when using Microsoft TV. This chapter uses the same content, as shown in Figure 11.1, to demonstrate how to create page backgrounds and image maps.

USING SUPPORTED IMAGE TYPES

Microsoft TV supports the following image formats:

- .gif
- .jpg
- .png

At this time, some graphics programs do not provide the ability to save images in the .png file format, so the included sample interactive TV pages use .gif and .jpg file formats exclusively.

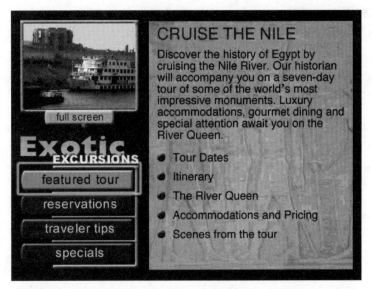

Figure 11.1 *The Exotic Excursions interactive TV content.*

ADDING IMAGES TO MICROSOFT TV CONTENT

The following sections describe easy-to-follow strategies for creating and adding images to content for Microsoft TV.

Image Width Should Not Exceed 544 Pixels

The Microsoft TV proxy server automatically resizes images wider than 544 pixels down to 544 pixels, unless the image is a background image for a document. The proxy server also resizes images, in some cases, if the total width of HTML elements, including images, exceeds 544 pixels. Therefore, when designing Microsoft TV content, it is best to keep the width of an image less than 544 pixels.

Image Colors Should Be Made NTSC-safe

The Microsoft TV proxy server automatically remaps colors so they are safely in accord with the requirements of the National Television Standards Committee (NTSC), but it is best not to rely on the proxy server to select colors for you.

For a full explanation of how to create images with NTSC-safe colors, see Chapter 9, "Selecting Colors for Microsoft TV Content."

Embedded Text Should Be Sans Serif Font, 16 Points or Larger

Microsoft TV uses Helvetica as the default font; when designing images, it is recommended that you use sans serif fonts that work well with Helvetica. Also, text embedded in images should be 16 points or larger to ensure readability from a distance.

Image Byte Count Should Be Minimized

When designing pages for interactive TV, strive to create visually appealing pages that load quickly. As a general rule, the samples in this book are each designed to be 20 KB or less, overall. To help keep byte count to a minimum for interactive TV pages, we recommend avoiding the use of gratuitous images and optimizing image file size. For example, for .jpg images, you can compress the image file size by reducing the image quality. For .gif files, you can reduce image file size by reducing the number of colors. For best results, we recommend that you experiment with images and display them on TV to get a feel for how far image size can be compressed.

Single-Pixel Lines Should Be Avoided

When creating graphic images, avoid using single-pixel lines, because they vibrate when shown on TV. For best results, make lines at least two pixels thick. Three pixels or more is recommended.

CREATING A USER INTERFACE WITH IMAGES AND JAVASCRIPT

An easy and effective way to build a dynamic user interface for interactive TV content is to use image swapping to provide user feedback. To demonstrate one way of swapping images when creating content for Microsoft TV, let's take a look at how the exotic_tv.html page of Exotic Excursions is constructed, as shown in Figure 11.2.

Many Web pages use image swapping for rollovers—they swap images when the mouse pointer is moved over or off of a particular image. For interactive TV content, however, rollovers seem somewhat redundant because the image with the focus is already well indicated by a large yellow box. A better use of image swapping is to change images to indicate that a particular button has been selected.

For example, the exotic_tv.html page contains button images that are swapped in response to a user selection. When the TV viewer clicks a button, the selected button changes to lavender to indicate it is selected, while the previously selected button changes to purple to indicate that it is no longer selected. The following sections explain how this user interface is implemented, including precaching images and swapping images with JavaScript.

Figure 11.2 *The exotic_tv.html page shown on the computer at design time.*

The Code for exotic_tv.html

The exotic_tv.html page uses JavaScript to preload images and to dynamically swap them based on a user selection. This section shows the code for the exotic_tv.html page and then dissects the code to show how various portions of it work. Here is the HTML code used to build the exotic_tv.html page:

```
<html>
<head>
<script type/text="JavaScript">
//creates image objects for swappable images
var featuredoff = new Image(190,42)
var featuredon = new Image(190,42)
var reservationsoff = new Image(190,38)
var reservationson = new Image(190,38)
var tipsoff = new Image(190,38)
var tipson = new Image(190,38)
var specialsoff = new Image(190,43)
var specialson = new Image(190,43)

//sets the src for image objects
featuredoff.src = "images/feature.gif"
featuredon.src = "images/feature2.gif"
reservationsoff.src = "images/reser.gif"
reservationson.src = "images/reser2.gif"
tipsoff.src = "images/tips.gif"
```

(continued)

```
tipson.src = "images/tips2.gif"
specialsoff.src = "images/specials.gif"
specialson.src = "images/specials2.gif"

//oldimage keeps track of the last button clicked
//oldsrc holds the off-state image for the object
var oldimage = ""
var oldsrc = ""

function featureload()
{
   parent.frames["content"].location.href="exotic_feature1.html";
   document.btn_feature.src = featuredon.src;
   if (oldimage != document.btn_feature)
   {
      oldimage.src = oldsrc
   }
   oldimage = document.btn_feature
   oldsrc = featuredoff.src
}

function reservationsload()
{
   parent.frames["content"].location.href="exotic_reservations1.html";
   document.btn_reservations.src = reservationson.src
   if (oldimage != document.btn_reservations)
   {
      oldimage.src = oldsrc
   }
   oldimage = document.btn_reservations
   oldsrc = reservationsoff.src
}

function tipsload()
{
   parent.frames["content"].location.href="exotic_traveler1.html";
   document.btn_tips.src = tipson.src
   if (oldimage != document.btn_tips)
   {
      oldimage.src = oldsrc
   }
   oldimage = document.btn_tips
   oldsrc = tipsoff.src
}

function specialsload()
{
   parent.frames["content"].location.href="exotic_specials1.html";
   document.btn_specials.src = specialson.src
```

```
    if (oldimage != document.btn_specials)
    {
        oldimage.src = oldsrc
    }
    oldimage = document.btn_specials
    oldsrc = specialsoff.src
}

</script>

<title>Exotic Excursions</title>
</head>

<body bgcolor="black">

<div style="position:absolute;top:14;left:14">
<img src="images/border.gif" width=190 height=145 alt="" border="0">
</div>

<div style="position:absolute;top:19;left:19">
<object data="tv:" width=180 height=135></object>
</div>

<div style="position:absolute;top:159;left:52">
<a href="tv:"><img src="images/full.gif" width=114 height=19 alt=""
border="0"></a>
</div>

<div style="position:absolute;top:194;left:14">
<img src="images/exotic.gif" width=190 height=47 alt="" border="0">
</div>

<div style="position:absolute;top:241;left:14">
<a href="JavaScript:onClick=featureload()">
<img src="images/feature2.gif" width=190 height=42 alt="" border="0"
name="btn_feature"></a>
</div>

<div style="position:absolute;top:283;left:14">
<a href="JavaScript:onClick=reservationsload()">
<img src="images/reser.gif" width=190 height=38 alt="" border="0"
name="btn_reservations"></a>
</div>

<div style="position:absolute;top:321;left:14">
<a href="JavaScript:onClick=tipsload()">
<img src="images/tips.gif" width=190 height=38 alt="" border="0"
name="btn_tips"></a>
</div>
```

(continued)

```
<div style="position:absolute;top:359;left:14">
<a href="JavaScript:onClick=specialsload()" name="specials">
<img src="images/specials.gif" width=190 height=43 alt="" border="0"
name="btn_specials"></a>
</div>

<script type/text="JavaScript">
oldimage = document.btn_feature
oldsrc = featuredoff.src
</script>

</body>
</html>
```

Preloading Images

When using image swapping to create a user interface, you can preload images into the TV receiver's cache so the TV viewer does not have to wait for the replacement images to load. Preloading, also called precaching, requires you to create an image object in memory and then assign a URL to the src property of the object, as shown in the following code snippet:

```
<script type/text="JavaScript">
imageone = new Image(190,42)
imageone.src = "feature.gif"
```

As shown in the preceding code, the "imageone = new Image" syntax creates an imageone object in memory. Also notice that width and height (190,42) properties are specified for the image object. Specifying the width and height property of the image object speeds load time. The width and height of the replacement image must match the width and height of the original image. Once the image object is created, you can dynamically assign the URL of an image to it, as demonstrated with the 'imageone.src = "feature.gif"' syntax.

You can preload images either in immediate script statements that run as the page loads or in response to the window's *onLoad* event handler. The following code shows how the "new Image" constructor is used to create eight image objects in memory. In this case, each button area for the Exotic Excursions content has two images, one for an "on" state and one for an "off" state.

```
<html>
<head>
<script type/text="JavaScript">
//creates image objects for swappable images
var featuredoff = new Image(190,42)
var featuredon = new Image(190,42)
```

```
var reservationsoff = new Image(190,38)
var reservationson = new Image(190,38)
var tipsoff = new Image(190,38)
var tipson = new Image(190,38)
var specialsoff = new Image(190,43)
var specialson = new Image(190,43)
```

Next, the src property of the objects is set to the URL of the images, as shown in the following code:

```
/sets the src for image objects
featuredoff.src = "images/feature.gif"
featuredon.src = "images/feature2.gif"
reservationsoff.src = "images/reser.gif"
reservationson.src = "images/reser2.gif"
tipsoff.src = "images/tips.gif"
tipson.src = "images/tips2.gif"
specialsoff.src = "images/specials.gif"
specialson.src = "images/specials2.gif"
```

When JavaScript sees a statement assigning a URL to an image object's src property, it instructs the browser to load the image into the cache. The preloading method covered in this section demonstrates just one way to preload images. There are several other methods you can use to preload images, depending on the needs of the content you are creating. You can find more information about preloading images in a variety of JavaScript books that are currently on the market.

> **CAUTION** Preloading images on the same page with the TV object can sometimes cause an image-load failure when a Web page first loads on a WebTV Plus Receiver. If your content preloads images, be sure to test the content thoroughly before putting it into production.

Making an Image Clickable

The specification produced by the Advanced Television Enhancement Forum (ATVEF) defines JavaScript 1.1 as the scripting language for creating interactive TV content. The JavaScript 1.1 object model, however, does not support the *onClick* event for the image object. Fortunately, there is an easy alternative—surround an image with an anchor, and then define an *onClick* event for the anchor. The following example shows how this is done for the feature2.gif image of the exotic_tv.html page:

```
<div style="position:absolute;top:241;left:14">
<a href="JavaScript:onClick=featureload()">
<img src="images/feature2.gif" width=190 height=42 alt="" border="0"
name="btn_feature"></a>
</div>
```

Referencing an Image

Because a document can have one or more images, image object references are stored in the JavaScript 1.1 object model as an array. You can reference an image by array index or by image name. For example:

```
document.images[n]
document.image_name
```

Before you can reference an image object by name, however, you must first assign a name to the image. For example, the feature2.gif image is given the name "btn_feature," as shown in bold in the following code:

```
<div style="position:absolute;top:241;left:14">
<a href="JavaScript:onClick=featureload()">
<img src="images/feature2.gif" width=190 height=42 alt="" border="0"
name="btn_feature"></a>
</div>
```

Creating Functions to Swap Images

To handle the image-swapping task when the user clicks an image, you must create a JavaScript function that is called by the *onClick* event of the anchor tag surrounding the image, as shown in bold in the following code sample:

```
<div style="position:absolute;top:241;left:14">
<a href="JavaScript:onClick=featureload()">
<img src="images/feature2.gif" width=190 height=42 alt="" border="0"
name="btn_feature"></a>
</div>
```

The featureload function is added to the <script> section of the document. As you can see in the following code sample, the featureload function first loads a new page into the content frame. The function then sets the src property of the image to the URL of the featureon.gif image. If the previously selected button was not the featured tour button, then the function swaps the previously selected button to the off state. It then stores the object and its src in the oldimage and oldsrc global variables.

```
function featureload()
{
    parent.frames["content"].location.href="exotic_feature1.html";
    document.btn_feature.src = featuredon.src;
    if (oldimage != document.btn_feature)
    {
        oldimage.src = oldsrc
    }
    oldimage = document.btn_feature
    oldsrc = featuredoff.src
}
```

To set the values of the oldimage and oldsrc variables to the default button state for the page when it first loads, the following script is added to the bottom of the document:

```
<script type/text="JavaScript">
oldimage = document.btn_feature
oldsrc = featuredoff.src
</script>
```

ADDING IMAGES AS BACKGROUND

When designing interactive TV content, images can be used as backgrounds for documents, divisions, or tables. For the content pages of the Exotic Excursions content, for example, an image is defined as the background of a division, as shown in Figure 11.3.

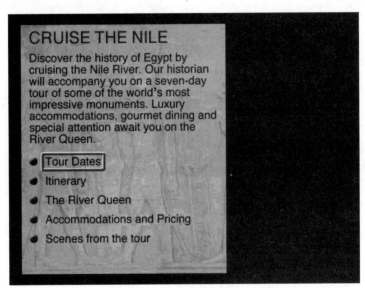

Figure 11.3 *The background image for the exotic_feature1.html page.*

Keep in mind that any images used as a background should be kept as small as possible to avoid delays in loading the page. The background for this sample is 4 KB. The following sample code shows how the background is added to the exotic_feature1.html page. Notice that the background image is specified here as a background-image attribute of the style tag.

```
<body bgcolor="black">
<div style="position:absolute;top:16;left:11;width:327;height:388;
background-image:url(images/back.jpg)">
</div>
```

USING IMAGE MAPS

Microsoft TV supports the use of client-side and server-side image maps. Server-side image maps, however, are not recommended for Microsoft TV content because they require the user to first select the entire image map, and then manually move a mouse pointer over the desired link. Client-side image maps, on the other hand, function quite nicely for Microsoft TV content. Figure 11.4 shows a client-side image map for the exotic_scenes1.html page of the Exotic Excursions interactive TV content.

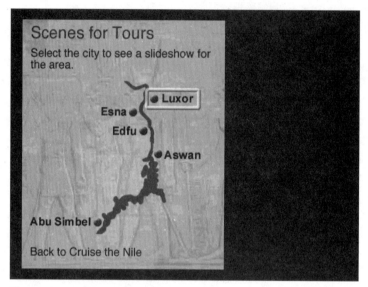

Figure 11.4 *An image map for the exotic_scenes1.html page.*

The following bold code shows how the image map is created for the exotic_scenes1.html page:

```
<html>
<head>

<link rel="stylesheet" type="text/css" href="cssTV.css">

<script type/text="JavaScript">
var bName = navigator.appName
if (bName == "Microsoft Internet Explorer")
{
    document.write("<link rel='stylesheet' type='text/css'
      href='cssPC.css'>");
}
```

```
</script>
<title>Featured Tour Content</title>
</head>

<body bgcolor="black">
<div style="position:absolute;top:16;left:11;width:327;height:388;
background-image:url(images/back.jpg)">
</div>

<div style="position:absolute;top:25;left:24;">
<font class="clsTitle">Scenes for Tours</font>
</div>

<div style="position:absolute;top:60;left:24;width:305">
<font class="clsDescription">Select the city to see a slideshow for the
area.</font>
</div>

<div style="position:absolute;top:110;left:24;width:305">
<img src="images/map.gif" width=278 height=242 alt="" border="0"
usemap="#citymap">
</div>

<div style="position:absolute;top:370;left:24;width:305">
<a href="exotic_feature1.html" target="content">Back to Cruise the Nile</a>
</div>

<map name="citymap">
<area shape="rect" coords="0,209,120,232" href="exotic_abu.html"
target="content">
<area shape="rect" coords="201,105,277,126" href="exotic_aswan.html"
target="content">
<area shape="rect" coords="130,65,194,92" href="exotic_edfu.html"
target="content">
<area shape="rect" coords="112,40,176,58" href="exotic_esna.html"
target="content">
<area shape="rect" coords="193,18,268,38" href="exotic_luxor.html"
target="content">
</map>

</body>
</html>
```

NOTE   Many HTML editors, such as Microsoft FrontPage and Macromedia Dreamweaver, provide built-in tools to create client-side image maps. See your HTML editor's online Help for more details about creating client-side image maps.

WHAT'S NEXT

This chapter explained how to use JavaScript to preload images and dynamically swap them. It also explained how to make an image a clickable object and how to call a function when an image is clicked. In addition, this chapter demonstrated how you can use image backgrounds and client-side image maps in your interactive TV content. In the next chapter, we'll take a look at how you can add even more visual appeal to your interactive TV content by adding animation to it.

Chapter 12

Adding Animation to Microsoft TV Content

In This Chapter

- Where to Find Sample Content for This Chapter
- Using Animated .Gifs
- Using DHTML to Create Animation
- Adding Macromedia Flash Movies

Animation adds entertainment value to interactive TV content. With animation, you can add pizzazz to a company logo, create text that flies across the screen, or visually demonstrate how a product works. In some cases, you may choose to use animation for the user interface of your content. For example, you might create animated buttons to solicit user feedback.

Microsoft TV supports three basic methods of adding animation to interactive TV content:

- **Animated .Gifs**—These are an easy way to add animation to a page.

- **Dynamic HTML (DHTML)**—This method lets you hide and show elements on a page, and lets you dynamically move elements to create special effects such as flying text.

- **Macromedia Flash Movies**—This technology can be used for product demonstrations, for special effects such as fading in a logo, and for creating interactive user interfaces. The Macromedia Flash Player is pre-installed on several of the standard interactive TV set-top boxes, including the WebTV Plus Receiver. However, Macromedia Flash is not considered a content creation standard by the Advanced Television Enhancement Forum (ATVEF), so it should be used only for content targeted at set-top boxes that contain the Flash Player.

This chapter discusses the three methods of adding animation to interactive TV content, describes the benefits and drawbacks of each method, and provides an example of how animation can be added to interactive TV content.

WHERE TO FIND SAMPLE CONTENT FOR THIS CHAPTER

This chapter uses the Tailspin Toys Animation sample content to demonstrate how to create animation for Microsoft TV. To get the most out of the material in this chapter, we recommend that you view the sample files on the companion CD and Web site.

- To view the Tailspin Toys Animation content on a set-top box running Microsoft TV or to view the content on the Web using your computer, go to *http://www.microsoft.com/tv/itvsamples*.

- To view the sample files associated with this chapter, see the "Creating Animation for Microsoft TV" topic in the "Microsoft TV Design Guide" section of the companion CD.

- For HTML source files of the Tailspin Toys Animation content, see the "Templates" section of the companion CD.

USING ANIMATED .GIFS

The easiest way to add animation to Microsoft TV content is to use an animated .gif file. An animated .gif file consists of a series of images that are swapped at various intervals to create a rudimentary form of animation. Adding an animated .gif file to a page is as simple as specifying an image with the HTML <img src="xxx.gif"> statement.

Large animated .gif images typically yield a high byte count, so it is best to use animated .gifs for small images such as corporate logos and ads. Figure 12.1 shows an animated .gif, located in the upper left corner of the Tailspin Toys page. The Tailspin Toys animated .gif loads the letters for the logo one letter at a time.

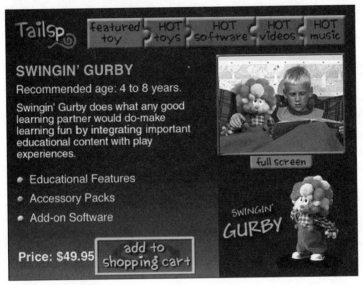

Figure 12.1 *An animated .gif used for the logo of Tailspin Toys.*

Creating Animated .Gifs

To create animated .gifs you can use any of the .gif animation programs available on the market, including several shareware products, such as Macromedia Fireworks, Adobe ImageReady, or Microsoft GIF Animator. While you can create animation with these programs, you must first create the artwork using a graphics program such as Adobe Photoshop.

For more information on creating an animated .gif, see the documentation with your graphics program and the .gif animator program.

USING DHTML TO CREATE ANIMATION

Dynamic HTML (DHTML)—the combination of HTML 4.0, scripting, and cascading style sheets (CSS)—can be used to create a wide variety of special effects. These include images or text that appear and disappear, images that float across the screen, and three-dimensional buttons that change their appearance based on user input. Many of these special effects rely on the *document.id.style* property, which is supported by Microsoft TV. For more information about using DHTML to create special effects, see Chapter 25, "DHTML for Microsoft TV."

ADDING MACROMEDIA FLASH MOVIES

Macromedia Flash, a software tool for creating vector-based animation, can be a good solution for producing animated interactive TV content, especially if you know that your target TV receivers have the necessary Flash Player installed. To play the Flash movies created with Macromedia Flash, the Flash Player must be installed on the TV receiver device. In most cases, the Flash Player must be pre-installed by the manufacturer of the TV receiver because most TV set-top boxes do not support downloading files. Macromedia Flash is considered an animation standard for Web content, but there is no guarantee that the Flash Player will be installed on all TV set-top boxes. At this time, the Microsoft TV platform does not license the Flash Player. The decision to include the Flash Player in a set-top box is made by individual set-top box manufacturers.

Typically, when you include a Flash file in a document, you encode it so that the user has the option of downloading the Flash Player if it does not exist on his or her hard drive. As the following example shows, you can also create code that loads a .gif image if the Flash Player is not installed. Macromedia provides a free utility called Aftershock that simplifies the process of adding a Flash movie to a page. Aftershock creates the JavaScript and HTML necessary to check for the Flash Player and, if not found, displays an alternate file instead. Aftershock is used for the animated logo added to the Tailspin Toys page, as shown in Figure 12.2.

Figure 12.2 *A Macromedia Flash movie added to Tailspin Toys.*

The following bold code shows how a Flash movie is added to the page. If a Flash Player is not detected on the receiving device, an alternate image, logo.gif, is displayed instead.

```
<body bgcolor="#0b0069">
<div style="position:absolute;top:14;left:12">
<!-- Aftershock logo.swf 3=113 4=49 6=1 18 19 -->
<object classid="clsid:D27CDB6E-AE6D-11cf-96B8-444553540000"
codebase="http://active.macromedia.com/flash/cabs/swflash.cab"
ID=logo width=113 height=49>
<param name=movie value="images/logo.swf">
<param name=loop value=false>
<param name=quality value=autohigh>
<param name=menu value=false>

<script language=JavaScript>
<!--
var ShockMode = 0

if (navigator.mimeTypes &&
    navigator.mimeTypes["application/x-shockwave-flash"] &&
    navigator.mimeTypes["application/x-shockwave-flash"].enabledPlugin)
{
    ShockMode = 1;
}

if (ShockMode)
{
    document.write('<embed src="images/logo.swf"');
    document.write(' width=113 height=49');
    document.write(' loop=false quality=autohigh menu=false');

    document.write('type="application/x-shockwave-flash" ⤸
          pluginspage="http://www.macromedia.com/ ⤸
          shockwave/download/index.cgi ⤸
          ?P1_Prod_Version=ShockwaveFlash">');

    document.write('</embed>');
}
else if (!(navigator.appName &&
          navigator.appName.indexOf("Netscape")>=0 &&
          navigator.appVersion.indexOf("2.")>=0))
{
    document.write('<img src="images/logo.gif" width=113 height=49 ⤸
          border=0>');
}

//-->
</script>
```

(continued)

```
<noembed>
<img src="images/logo.gif" width=113 height=49 border=0>
</noembed>
<noscript>
<img src="images/logo.gif" width=113 height=49 border=0>
</noscript>
</object>
<!-- EndAftershock logo.swf -->
</div>

<div style="position:absolute;top:14;left:130;">
<img src="images/feature.gif" width=415 height=49 alt="" border=0
usemap="#toolbar" name="toolbar">
</div>

<map name="toolbar">
<area shape="rect" coords="345,0,414,48"
href="JavaScript:onclick=musicload()">
<area shape="rect" coords="269,0,346,47"
href="JavaScript:onclick=videoload()">
<area shape="rect" coords="157,0,269,48"
href="JavaScript:onclick=softwareload()">
<area shape="rect" coords="90,0,157,47"
href="JavaScript:onclick=toysload()">
<area shape="rect" coords="1,0,89,47"
href="JavaScript:onclick=featureload()">
</map>
</body>
</html>
```

For more information on using Macromedia Aftershock, see the Macromedia Web site at *http://www.macromedia.com.*

WHAT'S NEXT

This chapter explained how to add animation to content for Microsoft TV. This chapter also demonstrated how you can add a Macromedia Flash movie to your content. In the next chapter, we will take a look at how Microsoft TV handles navigation. We will also discuss ways that you can control the path of the selection box on a Web page to create a predictable user interface for the TV viewer.

Chapter 13

Handling Navigation for Microsoft TV Content

In This Chapter

- Where to Find Sample Content for This Chapter
- How Navigation Works for Microsoft TV
- Forms and the Selection Box

To design interactive TV content that is easy for TV viewers to use, it helps to understand Microsoft TV's navigation model. With Microsoft TV, television viewers navigate between links and form elements using a remote control or an infrared keyboard. Unlike traditional Web pages, where users can move the mouse pointer freely about the page, Microsoft TV uses a tabbed navigation model. With the tabbed navigation model, TV viewers click directional arrow keys on the remote or keyboard to move in sequential fashion through the links and form elements on the page.

This chapter provides a detailed explanation of how navigation works for Microsoft TV, shows how you can set the initial focus for a Web page, and shows how you can control the logical movement of the selection box.

WHERE TO FIND SAMPLE CONTENT FOR THIS CHAPTER

This chapter uses the Exotic Excursions sample content to demonstrate how navigation works for Microsoft TV. To get the most out of the material in this chapter, it is recommended that you view the sample files on the *Building Interactive Entertainment and E-Commerce Content for Microscoft TV* companion CD and Web site.

- To view the Exotic Excursions content on a set-top box running Microsoft TV or to view the content on the Web using your personal computer, go to *http://www.microsoft.com/tv/itvsamples*.

- To view the sample files associated with this chapter, see the "Handling Navigation for Microsoft TV Content" topic in the "Microsoft TV Design Guide" section of the companion CD.

- For HTML source files of the Exotic Excursions content, see the "Templates" section of the companion CD.

HOW NAVIGATION WORKS FOR MICROSOFT TV

With Microsoft TV's tabbed navigation model, TV viewers click through links and form elements in a sequential fashion. When the TV viewer clicks a directional arrow on his remote control or keyboard, the selection box moves to the next closest link or form element in the direction of the arrow button. As shown in Figure 13.1, the selection box surrounds the featured tour link.

Now let's assume the TV viewer presses the Down Arrow key. In this case, the yellow selection box moves to the next available link or form element below it. By the same token, if the TV viewer clicks the Right Arrow key, the yellow selection box moves to the closest link or form element to the right, as shown in Figure 13.2.

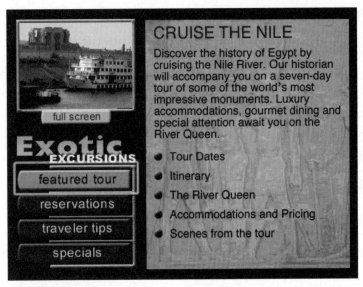

Figure 13.1 *The yellow selection box indicates the focus.*

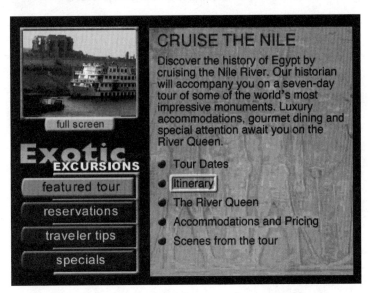

Figure 13.2 *The yellow selection box moved to a new link.*

How Navigation Works for Image Maps

With client-side image maps, the TV viewer navigates through the defined links on the image in the same fashion he or she would navigate through traditional links. However, server-side image maps operate somewhat differently. With server-side image maps, the TV viewer must first select the entire image, then use the directional keys on a remote or keyboard to move a yellow pointer over a link. Once the pointer is positioned over the link, the viewer can select it. The process is hardly intuitive. As a result, server-side image maps should be avoided for Microsoft TV content.

FORMS AND THE SELECTION BOX

Microsoft TV provides several special attributes to handle how the selection box works with form elements such as <select> and <input type=text>. The use of these special attributes is covered in more detail in Chapter 17, "Creating Forms for Microsoft TV Content."

Controlling Where the Selection Box Appears

In some cases, you might want to control where the selection box appears when a page loads. To do this, Microsoft TV provides a special selected property for the <a> tag. For example, by default, the yellow selection box appears on the Full Screen button because it is the first anchor in the page, as shown in Figure 13.3.

Figure 13.3 *The yellow selection box appears on the first available link.*

However, you might want to set the yellow selection box somewhere more positive than at the Full Screen button, which encourages the TV viewer to return to watching full-screen TV. A better initial location for the selection box might be the Tour Dates link, which encourages the TV viewer to explore the interactive content, as shown in Figure 13.4.

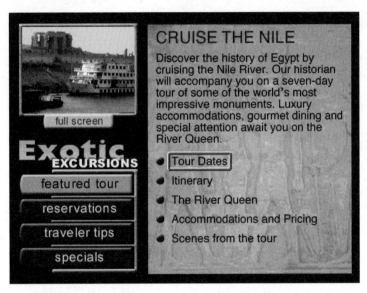

Figure 13.4 *The yellow selection box indicates a hyperlink to Tour Dates.*

To set the focus on the Tour Dates link when the page first loads, you add the selected property to the <a> tag as shown in bold in the following code example:

```
<div style="position:absolute;top:220;left:24">
<img src="images/bullet.gif" width=13 height=13 alt=""
border="0">  
<a href="exotic_sample1.html" selected>
Tour Dates</a>
</div>
```

Controlling the Direction of the Selection Box

For some interactive TV content, you might want to control the direction of the selection box. For example, pressing the Right Arrow key on the remote moves the selection box to the next closest selectable area that is to the right and down. In some cases, the next closest selectable area may be down, rather than to the right as preferred. To correct this situation and produce predictable navigation for the TV viewer, you can add an attribute to the <a> tag to specify where the selection box moves to when one of the four directional arrows is pressed.

To specify the direction of the selection box, you must first identify the target link—that is, the link that you wish the selection box to move to when a particular arrow key is pressed. To do this, you specify an i.d. for the link, as shown in the following sample code:

```
<a href="sample1.html" ID="sample1_link">Sample Page One</a>
```

For the link from which you want to control the direction of the selection box, you specify the i.d. of the target link as an attribute of the nextright, nextleft, nextup, or nextdown property. For example, the following code shows the target sample1_link defined for the nextright property.

```
<a href="tours1.html" nextright="sample1_link">Tours</a>
```

WHAT'S NEXT

This chapter explained how navigation works for Microsoft TV, and then demonstrated some useful techniques for building interactive TV content that is easy to navigate. The next chapter takes a look at Microsoft TV's support for audio and video.

Chapter 14

Audio and Video Support for Microsoft TV

In This Chapter

■ Supported Audio Formats

■ Supported Video Formats

■ Volume Control of the TV Object

Microsoft TV supports a variety of sound formats and two video formats at the time of this writing. The use of streaming media with Microsoft TV is in its infancy. Audio streaming is supported, but video streaming is not. This chapter lists the supported audio and video formats, but it does not provide any details about implementing them.

SUPPORTED AUDIO FORMATS

Microsoft TV supports the following audio formats:

- **AIFF**—Uncompressed format for 8- and 16-bit linear, A-LAW, u-LAW, IMA ADPCM compressed, and "none" formats for compressed AIFF files. File types include .afc, .aif, .aifc, and .aiff.

- **Basic audio** (usually Mu-LAW)—8- and 16-bit linear, A-LAW, u-LAW, and Adaptive Digital Pulse Code Modulation (ADPCM) format audio files from Sun Microsystems, Digital Equipment Corporation, and NeXT Software (now owned by Apple Computer). File types include .au and .snd.

- **Global System for Mobile Communications (GSM)**—voice compression scheme used for the GSM international digital cellular phone standard (8 kilohertz sampling rate). The file type is .gsm.

- **Macromedia Flash**—Version 2 format. The file type is .swf.

- **Musical Instrument Digital Interface (MIDI)**—General MIDI format and MIDI files with lyric tracks (also known as karaoke MIDI). File types include .kar, .mid, and .midi.

- **MOD**—Sampled music track files in most of the popular formats, including 669Mod, Protracker 15, Fastracker, Multitracker, Streamtracker, Ultratracker, and Unitracker. File types include .669, .mod, .mtm, .s3m, .stm, and .xm.

- **Moving Pictures Experts Group (MPEG)**—all versions of MPEG-1 including Layers I, II, and III, as well as the Low Sampling Frequency (LSF) versions of MPEG-II, including Layers I, II, and III. File types: .m1a, .m3u, .mp2, .mp3, .mpa, and .mpg.

- **Apple QuickTime**—streaming audio as well as movie soundtracks in the common uncompressed formats. File types include .mov and .qt, up to version 2.0.

- **RealAudio**—RealNetworks' RealAudio versions 1.0, 2.0, and 3.0. File types include .ra and .ram. Content must be streamed off a ".pnm" Real Server.

- **Shockwave**—Audio Shockwave files. The file type is .swa.

- **RMF**—Headspace's rich MIDI format music files and the Headspace Beatnik plug-in (supported on the WebTV Plus system only).

- **WAVE**—Microsoft Windows WAVE audio; 8- and 16-bit linear, A-LAW, u-LAW, IMA/DVI ADPCM, as well as ADPCM compressed formats. The file type is .wav.

  NOTE Any audio formats not listed above are not supported by Microsoft TV at this time.

SUPPORTED VIDEO FORMATS

At the time of this writing, Microsoft TV supports the following video formats:

- Moving Pictures Experts Group 1 (MPEG-1)
- VideoFlash

  NOTE Any video formats not listed above are not supported by Microsoft TV at this time.

VOLUME CONTROL OF THE TV OBJECT

With Microsoft TV, the volume of the TV object cannot be controlled separately from the TV volume. This means you can have two sound sources with only one volume control. As a result, it is best not to include the TV object on the same page as audio or video files.

WHAT'S NEXT

This chapter discussed the sound and video formats supported by Microsoft TV. The previous chapters in the "Microsoft TV Design Guide" section demonstrated how to create content for Microsoft TV. Now that we've explained how to create content, it is time to learn how to deliver it to the TV viewer. The following chapter, "Fundamentals of Delivering Interactive TV Content," discusses different methods you can use for delivering interactive TV content.

Part III

Delivering Microsoft TV Content

Chapter 15

Fundamentals of Delivering Interactive TV Content

In This Chapter

- Overview of ATVEF Transport Methods

- Transport A: Interactive TV Links

- Transport B: IP over VBI

- Transport A and Transport B Tradeoffs

- Interactive TV Vendors

Once you have created and completely tested interactive TV content, the next step is to deliver it to the TV viewer. The specification developed by the Advanced Television Enhancement Forum (ATVEF) defines two methods for delivering interactive

TV content: Transport A and Transport B. This chapter takes a look at both transport methods, discusses the relative merits of each method, and provides a list of interactive TV vendors who can handle the process of encoding data into the video signal so that it can be delivered to the TV viewer.

OVERVIEW OF ATVEF TRANSPORT METHODS

One of the key goals of the ATVEF was to make interactive TV content transport-agnostic. In other words, the ATVEF wanted to ensure that interactive TV content could be delivered over analog as well as digital television signals; over all standard broadcast modalities, including coaxial cable, satellite, and terrestrial broadcast; and over popular data networks of all types. The specification prepared by the ATVEF defines two methods for interactive TV content to be delivered using analog televisions signals: Transport A and Transport B. Both transport methods use triggers that may be sent through a part of the TV signal called the vertical blanking interval (VBI).

A Transport A trigger is a text string encoded into the video signal that causes a Go Interactive prompt to appear on the TV screen. A Transport A trigger can also pass JavaScript functions invisibly to Web pages on the TV receiver. Transport A triggers are strictly creatures of the analog world and, hence, always ride on the VBI highway. With the Transport A method, all data, including Web pages and their associated images, are downloaded through the back channel, which consists of a modem connected to a cable or phone line.

Transport B is more flexible than Transport A. Transport B carries both triggers and data in Internet Protocol (IP) packets. These Transport B packets may be carried over the analog VBI or over more modern digital routes such as a home satellite or a local area network (LAN). Either way, digital or analog, Transport B carries Web pages and their associated image files, along with information on how to use them, in IP packets. Transport B may or may not use a back channel, depending on the aim of the content developer.

Before we get into a more detailed discussion of Transport A and Transport B, let's take a look at the VBI and discuss its role in delivering interactive TV content.

NOTE   As digital television becomes universal, the need for VBI will be negated by higher-capacity digital signals.

What Is VBI?

With few exceptions, the television sets deployed in 1999 were analog. VBI is part of the analog video signal that does not constitute the viewable picture. Television pictures are made by drawing a series of vertically stacked horizontal lines that are scanned top to bottom, left to right. Each time a field of the television picture redraws, there are blank lines at the top of the screen available for the transport of data. Because National Television Standards Committee (NTSC) television pictures consist of two interlaced fields, two screen redraws are required to create each new television picture (referred to as a frame). To show fluid motion video, frames must refresh at a rate of almost 30 frames per second (fps). Anything less than 30 fps appears as a choppy image.

So what does VBI have to do with interactive TV? VBI is the portion of the video signal that is used for encoding interactive TV data, using either the Transport A or Transport B method.

TRANSPORT A: INTERACTIVE TV LINKS

Closed captioning for the hearing impaired is the most common example of data being transmitted over the VBI. Just as TV viewers need special set-top boxes like the Microsoft WebTV Plus Receiver in order to view interactive TV content, the hearing impaired community originally needed special equipment to decode closed captioning data on their television sets. Now most, if not all, modern televisions contain integrated circuitry that decodes closed captioning.

> **NOTE** In closed captioning, the words being spoken by television personalities appear as a stream of text at the bottom of the screen. Closed captions are turned on or off by a menu option built into the television. Before closed captioning, some programs featured open captioning, which is always visible. Some hearing viewers found open captioning objectionable and so closed captioning was created to solve this problem.

Now, in addition to closed captioning circuitry, some television manufacturers are beginning to include interactive TV decoders in their products. In fact, in the future, it is likely that interactive TV functionality will be a common feature built into most TV sets.

With the Transport A method of delivering interactive TV content, a string of text known as an interactive TV link is encoded into line 21 of the VBI, the same part of the VBI that is used for closed captioning. This text contains, among other things, the Uniform Resource Locator (URL) of a Web page. When the interactive TV link, encoded into the video signal, arrives at the TV receiver, the TV viewer sees a prompt that indicates interactive TV content is available. As shown in Figure 15.1, with a Microsoft TV–based set-top box the TV viewer can click the Go Interactive button to connect to the Internet and download the Web page specified in the interactive TV link.

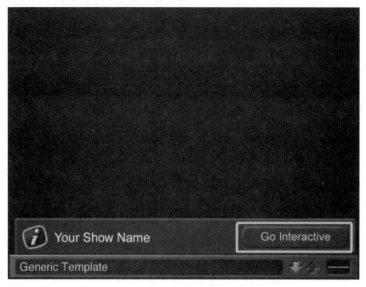

Figure 15.1 *The prompt that appears for an interactive TV link.*

NOTE   With interactive TV links, the functionality to display the Go Interactive prompt is built into the TV receiver.

The Web page is downloaded to the TV receiver via the back channel. For the WebTV Plus Receiver, the back channel is a standard 56-kilobits per second (Kbps) PC modem. However, some TV receivers have a digital subscriber line (DSL) or cable modem back channel capable of a higher data transfer rate than a standard PC modem.

VBI Line 21 Is Sacred

The Federal Communications Commission (FCC) has mandated that line 21 of the VBI must be set aside for closed captioning. Fortunately, closed captioning does not use all of the bandwidth available in line 21 of the VBI. In theory, most of the VBI lines remain available for data transmission. However, VBI lines are a very limited resource. In fact, many television stations have contractual obligations limiting the availability of VBI lines other than line 21.

The surety of having line 21 available is a primary reason that Transport A is such a compelling way to deliver ATVEF content. The Transport A method makes use of a part of line 21 to send Interactive TV links to ATVEF receivers without affecting closed captioning. Because the developers of closed captioning had foresight, they anticipated that need might arise for subtitled text in more than one language, or that additional space might be needed for information related to the television programming. As a result, the bandwidth in line 21 is separated into caption fields and text fields. The caption fields are strictly reserved for closed captioning. The text, or Text-2 (T-2), fields are used to transport interactive TV links. In a typical TV show, only about 25 percent of the bandwidth in line 21 is used by closed captioning. Very chatty shows may use as much as 75 percent of the bandwidth in line 21, but this still leaves plenty of room for interactive TV links, mainly because interactive TV links are so small.

What Are Interactive TV Links?

An interactive TV link is an ATVEF trigger transmitted on VBI line 21 using the T-2 service as specified in EIA-746a. In simpler terms, an interactive TV link is a text string that contains a URL, along with other information, and specifies a Web page with instructions for loading it. Because interactive TV links are encoded on VBI line 21 using the T-2 service (as opposed to the closed captioning service), interactive TV links can be sent without conflicting with closed captioning. Once a Web page is resident on the TV receiver, an interactive TV link containing a script trigger may be sent to initiate a function on the page. We will discuss script triggers in more detail in the following chapter. For now, let's take a look at a stripped-down interactive TV link, as shown in Figure 15.2.

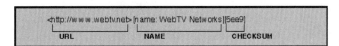

Figure 15.2 *A sample interactive TV link.*

For an interactive TV link, URL, type, and checksum attributes are required. The URL specifies the address of the Web page to download, while the type attribute, often called tve:, specifies the version of the ATVEF specification with which the link conforms. For example, [tve:1.1] would specify conformity with the specification produced by the ATVEF. The name attribute, as we'll discuss later in this chapter, specifies whether the interactive TV link appears to the TV viewer, or if it is a script trigger. The checksum attribute, as discussed in the following section, verifies the accuracy of the transmission.

Checksum Verifies Transmission Accuracy

All Hypertext Transfer Protocol (HTTP) connections are bi-directional. In bi-directional setups, when data is sent to a client, the client can send a message back to the server to indicate that the data arrived intact. Unlike two-way HTTP connections, the protocol for VBI is one way, which means there is no error checking in VBI transport. To compensate for the inability to check the delivery of correct trigger strings, Transport A uses a checksum attribute, which is discussed in more detail in the following section.

Using the WebTV Viewer to Create Interactive TV Links

Perhaps the easiest way to create an interactive TV link is to use the Microsoft WebTV Viewer. The WebTV Viewer contains an Interactive TV Link Creator window that automates the process of creating interactive TV links.

> **NOTE** If you have not yet installed the WebTV Viewer, see the section "Setup for the WebTV Viewer" of Chapter 3, "What You Need to Create and Deliver Microsoft TV," for downloading and installation instructions.

To open the Interactive TV Link Creator window (Figure 15.3), open the WebTV Viewer, and from the TV menu, click Show Link Creator Window.

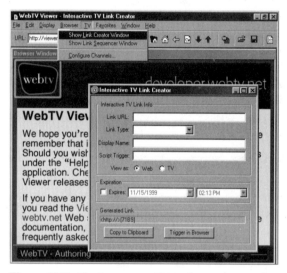

Figure 15.3 *The Interactive TV Link Creator window.*

In the Generated Link text box in Figure 15.3, notice the [71B9] value. This is the checksum value that is automatically calculated for you. The checksum value is expressed as a 16-bit number in hexadecimal notation. If you key a URL value into the Link URL text box, the checksum value in the Generated Link text box will change accordingly.

As shown in Figure 15.4, specifying a URL of *http://www.microsoft.com*, a Display Name of "Your show name," and Web as the View As property yields the following interactive TV link:

<http://www.microsoft.com>[n:Your show name][8BOB]

Each character in the interactive TV link has a value, and the checksum allows the TV receiver to check these values to ensure that the link is transmitted intact. If

Figure 15.4 *An interactive TV link.*

the value of the elements of an interactive TV link do not match the checksum, the link is not displayed.

> **NOTE** The checksum uses a highly complicated formula to verify that bits are received as they were sent. To learn more about checksum, refer to the specification produced by the ATVEF or in IETF RFC 1071.

Script Triggers

The Display Name and Script Trigger text boxes, as seen in Figures 15.3 and 15.4, are used to set name and script attributes in a trigger. Remember, there are two kinds of triggers: a link trigger that causes a user prompt to appear and a script trigger that fires an action. Links and triggers always include a URL, but they may also include a human-readable name, an expiration date, and some JavaScript.

If the name and URL attributes are specified, a user prompt with the Go Interactive button appears. When the viewer clicks the button, the specified Web page is downloaded. Script triggers, which are designed to run a JavaScript function on a page, do not contain the name attribute. Therefore, if the name attribute is missing, the URL in the trigger matches the current URL of the Web page resident on the TV receiver, and the expiration date has not been reached, then a script is run on the page. The function that is run is determined by the value of the script attribute. For example, if a function named gofullscr() is on a page named humform.html, the following script trigger could be used to initiate it:

<http://itv.microsoft.com/humins/humform.html>[s:gofullscr()][d08c]

Because the trigger has no name attribute specified, it will initiate the specified Web page without displaying a user prompt.

Interactive TV Link-Encoding Hardware

Interactive TV links are combined with a standard video signal using a piece of hardware known as an encoder. The combined video and data signal (in the form of an interactive TV link) may be sent directly over the broadcast head end and eventually to TV viewers, or it may be encoded onto tape, as shown in Figure 15.5.

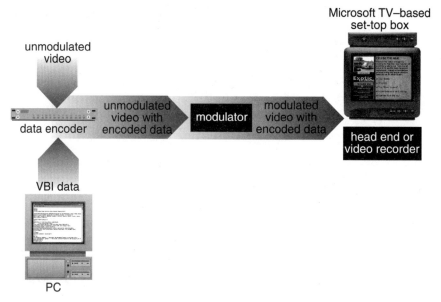

Figure 15.5 *Workflow to encode data into a video signal.*

If you plan to insert interactive TV links yourself, a data encoding machine is required, as shown in Figure 15.6.

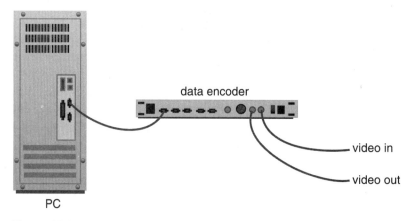

Figure 15.6 *Data encoder and data source computer.*

Figure 15.6 shows a typical setup for an encoder. Note that some encoders accept data through the serial COM port or through an Ethernet connection. Ethernet-enabled encoders are very handy for shops with multiple developers.

For more complete information on encoder vendors, see the "Interactive TV Vendors" section at the end of this chapter.

ATVEF Interactive TV Link Recommendations

Interactive TV links are time-tested, inexpensive, and flexible. When you create interactive TV links, the ATVEF committee recommends that you keep them as small as possible. In addition, the ATVEF recommends that you never send interactive TV links sooner than three minutes into a show; this prevents your data from stepping on the data of a previous show. The ATVEF also recommends that trigger transmissions never exceed 25 percent of the total bandwidth in Field 1 of VBI line 21. You can add interactive TV links to a live broadcast or in post-production. One important point to remember is that interactive TV links require a back channel (modem connection) to the Internet. If a back channel to the Internet does not exist, Transport B is your only choice.

TRANSPORT B: IP OVER VBI

With the Transport A method, a data string (trigger) sent over the VBI causes the bulk of the interactive TV content to load through the back channel. In Transport B, both the data and triggers are sent to the TV receiver as packets, using the proper protocol for their type through the same modality that brings in video signal. If the signal used to transmit video is analog, then VBI can carry the packets. If the modality is digital, the packets are carried as native format and no VBI is required. If your company controls the digital head end of a cable or satellite feed, or if your viewers are all on a LAN, then Transport B is your best transport method. In the analog world, three VBI lines are considered the practical minimum for Transport B, but you can sometimes get away with just one. In general, the more lines of VBI you have for sending data, the better off you will be.

What Is So Cool About Transport B?

Transport B, which sends data and triggers over the front channel, offers several important advantages over Transport A, which simply sends triggers. One advantage is that Transport B can be used to cache data, including HTML, JavaScript, image files, and other supporting material, ahead of the trigger that prompts the viewer to "Go Interactive." That way, when the viewer clicks the button, interactive data is already available on the TV receiver. As a result, the interactive content appears almost instantaneously. There's no waiting for the modem to connect and download the interactive TV content.

Another significant advantage of Transport B is that it prevents a flood of simultaneous hits to the server supplying the interactive TV content during popular, nationally televised TV shows.

NOTE   No back channel is required in Transport B, but the back channel is often used in the two-way communication required for e-commerce.

TCP/IP Multicasting Overview

Before you can clearly understand IP multicasting as it occurs in the Transport B method, you must understand a few basics about the Transmission Control Protocol/Internet Protocol (TCP/IP). TCP/IP is the *lingua franca* that makes the Internet, as well as most private networks, work. In order to be a node on the Internet, one of the things you do implicitly is agree to use the TCP/IP protocol.

When e-mail is sent or a Web page is served, the information is broken into pieces called packets. The TCP/IP protocol specifies how those packets are addressed. Conceivably, each packet sent can take a different path to the recipient. In addition, the packets can arrive out of sequence. As the packets arrive, they are reassembled into the original Web page or e-mail message by a group of programs called the TCP/IP stack.

In the context of interactive TV, you do not need to know what the members of the TCP/IP stack are or specifically what they do. However, if you are interested in how Announcements and Triggers use User Datagram Protocol (UDP) IP datagrams, while data flows in Unidirectional Hypertext Transfer Protocol (UHTTP), refer to the ATVEF specification and the book *TCP/IP Clearly Explained*, by Pete Lashin.

NOTE   UDP and UHTTP are required because of the unidirectional nature of even digital telecast media.

A range of addresses in the TCP/IP address system are called *multicast* addresses. By encoding IP packets of data into the lines of the VBI and then sending them to multicast addresses, you are multicasting. Multicasting over the VBI may be used to send both triggers and content in an analog television signal. Figure 15.7 shows a graphic representation of IP multicasting.

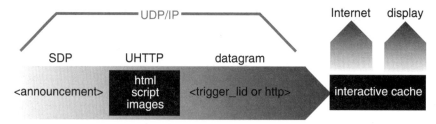

Figure 15.7 *IP multicasting has three main steps.*

The order of events in an IP multicast is as follows:

1. An announcement is sent telling the receiver what address or addresses within the multicast range to expect data to arrive on. The announcement also tells the receiver what size the local cache must be. Announcements use the Session Description Protocol (SDP) and a well-known address (224.0.1.113, port 2670). See the ATVEF specification for more details on SDP.

2. Data, including HTML files and their supported resources such as images, are compressed in GZIP format and sent as User Datagram Protocol (UDP) UDP/IP packets (datagrams) to the announced address using Unidirectional Hypertext Transfer Protocol (UHTTP). Web pages containing html, images, and scripting code are normally sent to the client-side interactive cache.

3. Triggers that look a lot like interactive TV links are sent, triggering some activity for the data that was previously sent down. Triggers may cause the Go Interactive prompt to appear or initiate a function on the current Web page. Triggers are sent in single User Datagram Protocol (UDP) UDP/IP multicast packets. This keeps triggers lightweight and nimble as well as adding a level of reliability much like checksum.

NOTE The interactive cache on a TV receiver can be as small as one megabyte. This number represents the high-water mark reached during a show, not the total content. If it is anticipated that content will come in just before it is needed, then it will be deleted after its usefulness has expired. The size of the content is measured as its decompressed value. ATVEF IP streams should be sent on the packet addresses 0x4b0 through 0x4bf. Other packet addresses may be used, but receivers are only required to handle IP datagrams arriving using packet addresses 0x4b0 through 0x4bf.

TCP/IP Protocol 101

All computers that communicate using the TCP/IP protocol agree on certain things. Every computer or node on a TCP/IP network must have a unique name and address. In fact, the whole Internet is really just one big TCP/IP network.

The address values on a TCP/IP network are four bytes in length and are represented as numbers separated by dots. The following numbers are the highest and lowest possible values on a TCP/IP network:

0.0.0.0
255.255.255.255

The numbers in IP addresses can be accurately thought of as 32-bit binary numbers. Thus, the numbers shown above are sometimes referred to as "all zeros" and "all ones," respectively.

NOTE   Neither 0.0.0.0 nor 255.255.255.255 may be used as proper IP addresses. They have special meaning in a TCP/IP network, but the explanation is beyond the scope of this book.

IP addresses for the Internet are assigned by a branch of the U.S. National Science Foundation called the Internet Network Information Center (InterNIC). Network addresses within an organization must begin with a one-, two-, or three-byte number, with remaining bytes expressed as zeros.

There were five types or classes of IP addresses created when the TCP/IP protocol was new. Four of them are shown in the following table; the fifth class of IP addresses is "reserved for future use" by InterNIC.

Class of Network	Address IP Range	Maximum Networks in Class	Maximum Hosts in Network
Class Type A	0.0.0.0 to 127.255.255.255	126	16 million plus
Class Type B	128.0.0.0 to 191.255.255.255	16,384	65,534
Class Type C	192.0.0.0 to 223.255.255.255	2,097,152	254
Class Type D	224.0.0.0 to 239.255.255.255	Reserved for multicasting	None

NOTE   You can determine what class of network you are part of by looking at the first three numbers (octet) returned when you key **ping servername** in an MS-DOS command prompt, where servername is the name of a server on your network.

The IP numbering scheme was created with different classes because it was thought that organizations of various sizes would have different needs. Many small businesses could function forever without needing more than 256 nodes, so, for them, a Class C network would work nicely. Other organizations might need between 256 and 65,534 nodes and benefit from Class B status. It was thought that only huge organizations such as the United States government would ever need Class A status.

How Announcements Get Heard

As discussed earlier in this chapter, the first step in multicasting is sending an announcement to tell the receiver what address or addresses within the multicast range to expect data to arrive on. But the packet containing the announcement is also data, so who announces the announcement?

The amazing thing about the announcement/multicasting scenario is the fact that Microsoft TV–based set-top boxes, and other TV receivers built to ATVEF guidelines, are always "listening" on certain multicast addresses for announcements.

The specification written by the ATVEF calls for announcements to be sent on a "well-known" address: 224.0.1.113 and port 2670. For the valid multicast addresses of target TV receivers, check with the TV receiver manufacturer or with the vendors of Transport B IP players. A list of interactive TV vendors is provided at the end of this chapter.

Local Identifier URL Scheme ("lid:")

Announcements and triggers come in two types: HTTP and lid. You're probably already familiar with URLs that begin with HTTP, but, most likely, URLs with the lid scheme are new to you. The lid scheme is used to funnel content into the interactive cache of the TV receiver. For example, if an announcement begins "<lid:\\myserver.com\somedir\somepage.html…," then the content is saved in the interactive cache. The receiver's interactive cache saves both the HTML pages and the directory structure. It is as if the whole Web site that makes up an interactive TV show is reconstructed in the interactive cache on the receiver.

HTTP Triggers

When a trigger begins with syntax such as "<http://www.someserver/somedir/somepage.html>," the TV receiver attempts first to find the page in the interactive cache. If the page specified in the announcement does not exist in the cache, then the TV receiver attempts to retrieve the page from the Internet through the back channel. If a back channel does not exist (i.e., there is no connection to the Internet), the page is not loaded onto the TV receiver.

Political Considerations of Using IP over the VBI

Currently, there is one primary "political" reality that makes Transport A a more realistic solution than Transport B in the analog space. Major broadcasters distribute television shows to their network of affiliate stations via satellite. In industry parlance, sending information via satellite is known as "squirting the bird." Because it is expensive to "squirt the bird," broadcasters typically strip VBI data (exclusive of closed captioning) out of their programs before sending them to the affiliates. Closed captioning is maintained by FCC fiat and therefore is safe from this stripping process. It is up to the affiliate stations to encode VBI data back into the signal before transmission to customers in their local area. Because there are hundreds of affiliate stations, all with various agreements about VBI usage, it is very difficult to use IP over VBI in practice.

Digital Video Signals

Although video encoders and the VBI provide a nice way to deliver content in today's predominantly analog world, digital transmission is also possible. Digital broadcast is definitely in the near future for the majority of the commercial television industry. When the world fully embraces digital television, the VBI will become just another anachronistic technological curiosity. One current example of digital delivery is the Moving Picture Experts Group (MPEG-2) stream sent to satellite set-top systems such as EchoStar.

The Trigger Receiver Object

Interactive TV HTML pages that will have triggers sent to them must have the HTML <object> tag to include a trigger receiver object on a page. The trigger receiver object processes triggers for the associated enhancement in the context of the page containing the object. The content type for this object is "application/tve-trigger." If a page consists of multiple frames, only one page may contain a receiver object. The following code snippet shows an example of the trigger receiver object:

```
<object type="application/tve-trigger"  id="objReceiverObj">
</object>
```

The page containing a trigger receiver should be at the top of the Document Object Model (DOM) for the interactive TV content. In a frames-based set of content, the trigger receiver object is placed in the frameset page, typically named something like base or main.html. JavaScript functions intended to be fired by triggers should also be placed in the main frameset page.

> **NOTE**   Some TV receiver manufacturers build the trigger receiver object functionality into the receiver. For example, content designed for the WebTV Plus Receiver does not require the trigger receiver object because the trigger receiver object is built into the hardware. However, for best results across the widest possible TV receivers, it is good practice to include the trigger receiver object on the top-level page in your interactive TV applications.

TRANSPORT A AND TRANSPORT B TRADEOFFS

Although Transport B has some decided advantages, Transport A is the current mode of choice for delivering content to ATVEF receivers. In the near future, the political issues inherent in VBI transport will disappear with the advent of relatively wide digital bandwidth. This will not be the end of the back channel, but it will probably obviate the need for Transport A. In the digital future, Transport B will bring in the interactive TV content, but using Transport B will still be a one-way street. When TV viewers actually want to buy something, the two-way communication provided by the back channel will still be the way to do it.

ATVEF receivers may choose to support only Transport B, IP-based trigger streams. They may also choose to support only interactive TV links if IP trigger streams are absent. But most TV receivers built to ATVEF guidelines will support both interactive TV links (Transport A) and IP trigger streams (Transport B) simultaneously. When a TV receiver supports both Transport A and Transport B simultaneously, ATVEF specifies the behavior described next.

When a broadcast data trigger is encountered, its URL is compared to the URL of the current page. If the URLs match and the trigger contains a script, the script will be executed. If the URLs match but there is no script, the trigger is considered a retransmission of the current page and will be ignored. If the URLs do not match and the trigger contains a name, the trigger is considered a new enhancement and will be offered to the viewer. If the URLs do not match and there is no name, the trigger will be ignored.

INTERACTIVE TV VENDORS

Norpak Corporation, Steeplechase Media, Screamingly Different Entertainment, and VITAC Corporation are given as examples of vendors you may wish to contact. However, *no endorsement or preference for any particular vendor should be inferred from its inclusion in this chapter.* The Microsoft TV Web site is a great resource for locating encoding companies as well as manufacturers of encoding equipment and may be found at the following URL:

http://www.microsoft.com/tv/interactive/resources/re_resources_01.asp

Encoder Vendors

Norpak Corporation is the world's leading supplier of TV data broadcast products, systems, and software for transmitting data over the VBI of any standard TV signal. Norpak delivers a complete line of TV data encoders, receivers, bridges, and system applications software for TV broadcasters, cable TV system operators, news agencies, system integrators, OEM suppliers, and value-added service providers. This company may be reached at:

Jim Carruthers, Ph.D
President, Norpak Corporation
10 Hearst Way
Kanata, ON K2L 2P4
Canada
Phone: (613) 592-4164
E-mail: *jimcarr@norpak.ca*
http://www.norpak.ca/

Norpak Encoders

■   TES3. For analog networks (most cases), order this model.

■   TES5. For digital head ends, order this model.

Encoders like the TES3 and TES5 made by Norpak are hardware, so in order to make them function you need software like that listed below.

Norpak Software

■   EIA-516 NABTS Data Broadcast Software. This module enables a TES3 or TES5 encoder to insert interactive content into a video signal on lines 10–20 of the VBI.

■   EIA-608 Caption Encoder Software. This module enables TES3 or TES5 closed captioning support. In more technical terms, it enables a TES3 or TES5 to insert data into line 21 of the VBI of a video signal.

Interactive TV Development Shops

If you'd rather contract the whole job of creating and encoding your interactive TV application, consider contacting Steeplechase Media or Screamingly Different Entertainment.

Steeplechase Media

Steeplechase Media is a leading new media production company with broad experience in both the entertainment and technology industries. It provides production services for interactive DVDs and interactive TV programming for Microsoft TV–based Internet Receivers. In creating interactive TV content, Steeplechase Media adheres to guidelines outlined in the specification created by the ATVEF. For more information, contact:

Rick Portin
Executive Vice President
Steeplechase Media, Inc.
201 Santa Monica Blvd
Suite 400
Santa Monica, CA 90401
Phone: (310) 451-0451
FAX: (310) 393-1147
E-mail: *rportin@steeplechase.net*

Screamingly Different Entertainment

Screamingly Different Entertainment provides interactive content to enhance television. Their services include visionary and creative ideas for TV program enhancements. For more information, contact:

3401 Winona Ave.
Burbank, CA 91504
Phone: (877) 262-6102 or (213) 680-3330
Fax: (316) 267-2472
E-mail: *info@screaminglydifferent.com*

Outsourcing Video Encoding

Interactive TV programmers commonly outsource the task of video data encoding. Most television affiliates will have video-encoding equipment in-house and may be willing to do it for you. Alternately, you may wish to subcontract the whole encoding process to people who do it every day. VITAC Corporation is the leading video encoding company in the United States. Chances are that the producer of your show is using VITAC for closed captioning already. As of this writing, these VITAC representatives are available to help you produce the very best interactive programming.

East Coast

Jeff Hutchins
Executive Vice President
101 Hillpointe Drive
Canonsburg, PA 15317
Phone: (800) 27-VITAC
E-mail: *jeff-h@vitac.com*
Bob Byer
Operations Manager
2030 M Street, NW
Suite 603A
Washington DC 20036
Phone: (202) 293-3707
E-mail: *b-b@vitac.com*

West Coast

Deborah Shuster
Vice President, Sales
4450 Lakeside Drive, Ste 250
Burbank, CA 94505
Phone: (888) LA-VITAC
E-mail: *deborah-s@vitac.com*

WHAT'S NEXT

The purpose of this chapter has been to provide you with a solid foundation in the fundamentals of delivering interactive TV content. Now it's time to take a look at the nuts and bolts of creating an interactive TV link, as described in the next chapter.

Chapter 16

Creating Interactive TV Links

In This Chapter

- A Review of Interactive TV Links

- Creating Interactive TV Links with the WebTV Viewer

- The Microsoft TV Interface for Interactive TV Links

- Creating a Link to Humongous Insurance

- The Link Type and View Attributes

- Trigger Expiration

- Creating Interactive TV Links with Script Triggers

- Interactive TV Link Sequencer

- Encoding Links into the VBI

- Trigger Syntax

Interactive TV links, as introduced in Chapter 15, "Fundamentals of Delivering Interactive TV Content," provide an easy and flexible way to deliver interactive TV content to TV receivers. Microsoft TV can utilize two types of interactive TV links: those with link triggers and those with script triggers. Interactive TV links with link triggers show the viewer a pop-up prompt to indicate that interactive TV content is available. Interactive TV links with script triggers are invisible to the TV viewer and fire a JavaScript function on a page loaded on the TV receiver. Regardless of the type of trigger, interactive TV links download Web content from the Internet through the modem-connected back channel.

This chapter explains how to create interactive TV links with link triggers and script triggers. As a teaching tool, we will use the Microsoft WebTV Viewer, which has an Interactive TV Link Creator that automates the process of creating interactive TV links. We will also demonstrate how to use a simple link utility to encode interactive TV links with a Norpak encoder. This utility is provided on the *Building Interactive Entertainment and E-Commerce Content for Microsoft TV* companion CD.

> **NOTE**   At the time this chapter was written, the WebTV 2.0 Viewer was the best simulator available for the creation of interactive TV links. Because it was designed and built before the adoption of the Advanced Television Enhancement Forum (ATVEF) Specification for Interactive Television 1.1, some of the screens generated by the 2.0 Viewer are not exactly like the ones you will see in Microsoft TV. This chapter identifies where and under which circumstances cosmetic differences between the 2.0 Viewer and Microsoft TV will occur. Most likely, a new Microsoft TV Viewer will be available in the near future. To check on its availability, see *http://www.microsoft.com/tv*.

A REVIEW OF INTERACTIVE TV LINKS

Most broadcast media define a way for text to be delivered along with the video signal. In some systems, this is called closed captioning or text-mode service. In other systems, it is called teletext or subtitling. The specification adopted by the ATVEF refers to triggers delivered via these mechanisms as broadcast data triggers or Transport A triggers. In general industry parlance, however, Transport A triggers are referred to as interactive TV links—which is what we will call them throughout this chapter.

Interactive TV Links in NTSC Television

In North America, the system used for broadcasting video signals has been standardized by the National Television Standards Committee (NTSC). In the NTSC system, which is the system that this book focuses on, text delivered with the video signal is called closed captioning. Interactive TV links, which are essentially text strings, are delivered along with closed caption text in a portion of the video signal called "line 21, text 2" of the vertical blanking interval (VBI). Developers who plan to send Transport A triggers in systems other than NTSC, such as EIA, ATSC, and DVB, should consult their transport documentation to learn how text is combined with video in their media.

Most of the world outside of North America has video transmission capability that can realistically carry User Datagram Protocol (UDP) packets for triggers, and Unidirectional Hypertext Transfer Protocol (UHTTP) packets for data (one-way multicasting). As a result, Transport B [Internet Protocol (IP) over VBI] is generally the preferred method of interactive TV content delivery outside of North America.

Keep in mind that interactive TV links are a solution devised to deliver digital data in the analog North American TV system. No doubt, in the near future, the analog transmission of video signals may become a historical curiosity. If you have control of a digital pipe, you will want to consider Transport B as your solution for delivering data to TV receivers, as explained in Chapter 15.

CREATING INTERACTIVE TV LINKS WITH THE WEBTV VIEWER

The WebTV 2.0 Viewer has a handy feature called the Interactive TV Link Creator, which enables users to create and test interactive TV links. This section shows how you can use this feature to make a simple interactive TV link.

If you have not downloaded and set up your WebTV Viewer, do so now using the directions supplied in Chapter 3, "What You Need to Create and Deliver Microsoft TV Content."

Setting Up the WebTV Viewer for Creating Interactive TV Links

Just like the real WebTV Plus Receiver, the WebTV Viewer has a Web mode and a TV mode. To simulate and test interactive TV links, you must put the WebTV Viewer in TV mode.

To put the WebTV Viewer in TV mode:

1. Open the WebTV Viewer on your personal computer.

2. If the remote control is not visible in the Viewer, click Remote Control on the Window menu.

3. On the remote control, click the View button. You should now see a window similar to the one displayed in Figure 16.1. In this figure, the TV test pattern with the "WebTV Networks, Inc." text represents the TV screen.

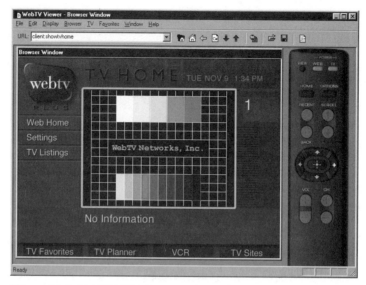

Figure 16.1 *The WebTV Viewer in TV mode.*

NOTE   You may wish to add custom backgrounds for your simulations by choosing Configure Channels from the TV menu and clicking Add on the Configure Channels dialog box to open the Edit Channel dialog box. To add a custom TV background image, choose Other from the Image drop-down list box on the Edit Image dialog box, and navigate to a prepared .jpeg file. This can be useful when you are demonstrating a concept to executives of a television network or station. Your demo will be more convincing if it has a background that includes a screen capture of on-air personalities or a show that potential clients will recognize.

Changing Channels and Viewing Full-Screen TV

To simulate an interactive TV link, you need to put the TV Viewer in full-screen TV mode.

To set the WebTV Viewer in full-screen TV mode:

As shown in Figure 16.1, a selection box surrounds the simulated TV screen. On the remote control, click the Go button (the oval-shaped button surrounded by arrow keys). You should now be in full-screen TV mode.

To change the channel:

On the remote control, put the mouse pointer over the "+" button (located under the CH text) and click it to choose a preset or custom channel. The Viewer shown in Figure 16.2 illustrates the use of a custom channel image.

Figure 16.2 *The WebTV Viewer with custom channel image.*

Creating an Interactive TV Link to a Standard Web Page

The simplest way to get started with interactive TV links is to create a link to a standard Web page on an existing Web site. In this case, a standard Web page is a Web page that does not have a TV object or any of the code associated with a TV object. At the time of this writing, Microsoft and a few other companies have television commercials that contain interactive TV links to their Web sites. As you will see in the following section, when you create a link to a standard Web page, Microsoft TV automatically places a small TV window at the bottom right corner of the page to maintain continuity with the TV programming.

> **NOTE**   Microsoft TV optimizes Web content for TV viewing, but in some cases Web pages designed for display on a personal computer may not look ideal on TV. For more information about optimizing Web content for TV viewing, see "Part II: Microsoft TV Design Guide" of this book.

To demonstrate how to create a link to a standard Web page, let's create a link to the Microsoft TV site at *http://www.microsoft.com/tv*. To do this, we'll use the Interactive TV Link Creator provided in the WebTV Viewer.

To open the Interactive TV Link Creator:

1. If your personal computer is currently not connected to the Internet, establish a connection now. You'll need the connection to test the interactive TV link, as described later in this chapter.

2. After connecting to the Internet, open the WebTV Viewer if it is not open already.

3. Make sure the TV screen is in full-screen mode and tuned to the channel of your choice, as described in the "Changing Channels and Viewing Full-Screen TV" section earlier in this chapter.

4. From the TV menu, click Show Link Creator Window. The Interactive TV Link Creator window will appear, as shown in Figure 16.3.

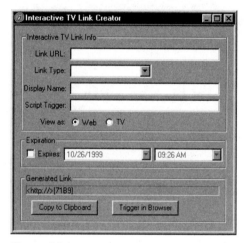

Figure 16.3 *The Interactive TV Link Creator.*

Creating a Link to *http://www.microsoft.com/tv*

To create a link in the Interactive TV Link Creator dialog box, specify a URL in the Link URL box and then type a name in the Display Name box. When you type a name in the Display Name box, it specifies that you want a prompt to appear to the TV viewer to indicate that interactive TV content is available. We will call this a visible trigger.

For this example, you create an interactive TV link with a visible trigger to *http://www.microsoft.com/tv*. When the link is received, a prompt appears to the TV viewer, who for this example is you.

NOTE Later in this chapter we will describe how to create interactive TV links with script triggers.

To create a link to *http://www.microsoft.com/tv*:

1. In the Link URL box of the Interactive TV Link Creator dialog box, type **http://www.microsoft.com/tv**.

2. In the Display Name box, type **Microsoft TV Website**. Leave all the values other than Link URL and Display Name in their default condition. The Interactive TV Link Creator dialog box should look like Figure 16.4. Be sure to leave the Interactive TV Link Creator open for the next set of steps.

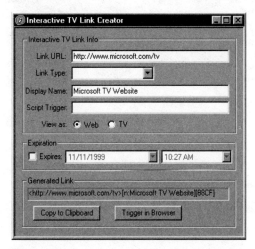

Figure 16.4 *An interactive TV link for* http://www.microsoft.com/tv.

As shown in the Generated Link box in Figure 16.4, the Interactive TV Link Creator automatically constructs the interactive TV link based on the values you specify:

```
<http://www.microsoft.com/tv>[n:Microsoft TV Website][88CF]
```

A more detailed explanation of the syntax of this link is provided later in this chapter, in the "Trigger Syntax" section. For now, let's look at how the link you created operates.

To trigger the interactive TV link in the WebTV Viewer:
In the Interactive TV Link Creator, click Trigger in Browser. An Interactivity icon will appear in the upper right corner of the screen, as shown in Figure 16.5.

Figure 16.5 *Simulated link in the WebTV Viewer.*

Notice the Interactivity icon in the upper right portion of the screen. In the WebTV Viewer, the response to a trigger arrival is different from what actually happens in Microsoft TV, but the behavior is essentially the same.

To go to *http://www.microsoft.com/tv*:

1. Click the Interactivity icon in the TV window. You will see a prompt, as shown in Figure 16.6.

2. Click Go to Web Page.

When you click Go to Web Page, the WebTV Viewer simulates what actually happens with an interactive TV link. That is, the URL specified in the link is compared to the URL in the cache, and then to the URL at the Web site. When a match is found, the page is downloaded to the TV receiver. You will then see the default page of *http://www.microsoft.com/tv*, as shown in Figure 16.7. Notice that a pop-up TV window is presented at the bottom right of the page to provide continuity with the television program.

The TV window, as shown in Figure 16.7, is the same window that users of a WebTV Plus Receiver will see if they choose TV Window from the Options dialog box while in Web mode. The benefit of having a pop-up TV window as part of an interactive TV link is that viewers who go to a sponsor's Web site will still hear and see the show they were watching previously. Moreover, no additional code is required to make this happen.

Figure 16.6 *The Go to Web Page prompt in the WebTV Viewer.*

Figure 16.7 *The* http://www.microsoft.com/tv *site.*

Simple links to existing content are a great way to get clients started with interactive TV content, but chances are that clients won't be satisfied for long. A much better solution is to create interactive TV links to Web content that is optimized for TV. We will explore interactive TV links to TV-optimized content in more detail in the following sections.

Creating Real Interactive TV Links

The Interactive TV Link Creator works great for generating the text string for an interactive TV link and for demonstrating how links work. Nevertheless, the question still remains: How do you encode the text string into the video signal? The answer is that there are a variety of third-party vendors who provide tools and services for encoding links. For a list, see the "Interactive TV Vendors" section in Chapter 15. In addition, the companion CD provides a handy Microsoft Visual Basic Scripting Edition application called the link utility.

THE MICROSOFT TV INTERFACE FOR INTERACTIVE TV LINKS

In Microsoft TV, a semitransparent icon does not appear in the upper right corner of the television screen. Instead, a pop-up prompt appears at the bottom of the screen, as shown in Figure 16.8.

Figure 16.8 *The Microsoft TV prompt for interactive TV links.*

In Figure 16.8, the "Microsoft TV Website" text is the value of the name attribute set in the interactive TV link [n:Microsoft TV Website]. This value is specified as the Display Name value in the Interactive TV Link Creator. When the TV viewer clicks the Go to Web Page button, a transition screen with a status bar appears as the TV receiver downloads the Web page from the Internet through the modem-connected back channel.

Interactive TV Link Limitations

The total number of characters you can transmit in an interactive TV link is limited both by the physical constraints of the VBI (two characters per frame and 30 frames per second) and by the encoder. However, interactive TV links are typically very small, so these size constraints are rarely a problem.

CREATING A LINK TO HUMONGOUS INSURANCE

The previous section demonstrated how to create a link to an existing Web site. In this section, we'll show how you can create an interactive TV link to TV-optimized content on a local Web server. In this way, you can test content and links locally before posting the content to an external Web server.

This section uses the Humongous Insurance demo on the companion CD. To run the examples covered in this section with the Humongous Insurance demo, your workstation must be connected to a computer running Internet Information Services (IIS) via your Local Area Network (LAN), or you must be running Personal Web Server (PWS) on a Microsoft Windows 98 computer.

> **NOTE** For more information about setting up an interactive TV environment, see Chapter 3, "What You Need to Create and Deliver Microsoft TV Content."

The steps provided in the following sections will work for any of the samples in the Templates folder on the companion CD. Of course, you'll need to change the names where necessary. What's more, the steps shown below should serve as a useful guideline for creating interactive TV links for your own content.

Copy the Humongous Insurance Demo to Your Server

To begin, copy a folder from the companion CD to the wwwroot folder of the Inetpub directory on the C drive of your IIS or PWS machine.

To copy the HumIns folder:

1. Place the companion CD in your drive bay and open the CD.

2. In the left-hand navigation panel, click Microsoft TV E-Commerce and then click Humongous Insurance.

3. At the bottom of the page, you'll find the HumIns folder. Copy the HumIns folder from the CD-ROM to the wwwroot folder of the Inetpub directory on the C drive of your IIS or PWS machine, as shown in Figure 16.9.

> **NOTE** The files in the HumIns folder are read-only. To modify the files, you must uncheck their Read-only attribute. To uncheck the Read-only attribute for the files, double-click the HumIns folder you just copied. From the Edit menu in Windows Explorer, click Select All. Right-click on a file in Windows Explorer, and then click Properties from the shortcut menu. Uncheck the Read-only option, and then click OK.

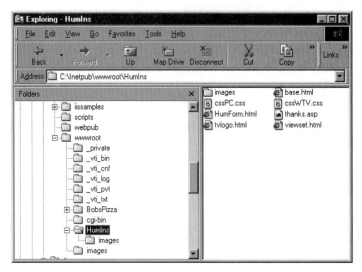

Figure 16.9 *Humongous Insurance example in a Web server directory.*

Determine the Name of Your Local Web Server

Before you can create a link to a local Web server, you must first determine that server's name. If you have the interactive TV setup recommended in Chapter 3, "What You Need to Create and Deliver Microsoft TV Content," you have one computer that serves as a development station and another computer connected to the LAN that functions as the Web server, as shown in Figure 16.10.

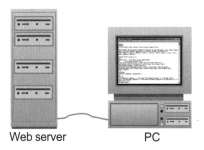

Web server   PC

Figure 16.10 *A Web server connected to a PC via a LAN.*

If you have the recommended workstation as shown in Figure 16.10, you can use the name of the computer plus the virtual root to IIS on that machine. The computers may be any combination of Microsoft Windows 98, Windows NT, or Windows 2000. IIS is used on machines running Windows NT and Windows 2000 Server. If the Web server is running Windows 98, it will run PWS.

If you are using a single Windows 98 machine as both your development machine and your Web server, use the computer's IP address. To find a computer's name, right-click on Network Neighborhood, click Properties, and look at the Computer name: text box on the Identification tab. To determine a lone machine's IP address, type **ipconfig** in an MS-DOS window as shown in Figure 16.11.

Figure 16.11 *Ipconfig returns the IP address of a lone computer on a LAN.*

Creating a Visible Link to Humongous Insurance

Now that you know the name of your Web server, let's create a link to the base.html page in the HumIns folder on that server. For this example, the interactive TV link is a visible link. Remember, a visible link is created by specifying a Display Name for the link, so that when the TV receiver gets the link, the TV viewer is prompted to load the interactive TV content.

To create an interactive TV link to base.html on your local server:

1. In the WebTV Viewer, open the Interactive TV Link Creator by clicking Show Link Creator Window from the TV menu.

2. In the Link URL box of the Interactive TV Link Creator dialog box, enter the URL ***http://servername/humins/base.html***, where servername is the name or IP address of your local server.

3. In the Display Name box, type **Humongous Insurance**. Leave all values other than Link URL and Display Name in their default condition. The Interactive TV Link Creator dialog box should look like the illustration shown in Figure 16.12.

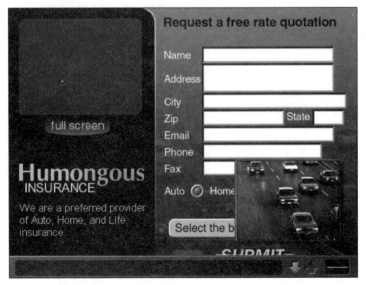

Figure 16.12 *An interactive TV link to base.html.*

4. Make sure the WebTV Viewer is in TV mode. If it is not, see the instructions earlier in this chapter under the heading "Setting Up the WebTV Viewer for Creating Interactive TV Links."

5. In the Interactive TV Link Creator, click Trigger in Browser.

An Interactivity icon appears in the upper right corner of the TV screen. Click the icon, and then click the Go to Web Page button. The base.html page of the Humongous Insurance demo will appear, as shown in Figure 16.13.

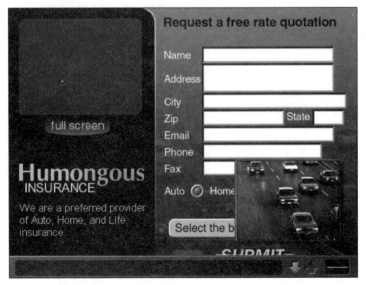

Figure 16.13 *Base.html page of Humongous Insurance.*

Can you see what is wrong with the way things are rendered in Figure 16.13? First, no video appears in the upper left corner of the screen (the area just above the Full Screen button). In addition, the built-in TV window appears in the bottom right corner of the screen. This situation occurs when the mode is not set properly for the interactive TV link. Microsoft TV has two modes: Web and TV. If you want video signal to appear in the TV object on a Web page, you must specify TV mode for the interactive TV link.

> **NOTE** Each of the sample demos on the companion CD contains a viewset.html page that simulates an interactive TV link by opening the page in TV mode, as shown in the following code snippet:
>
> ```
> <a href="base.html" view="tv"><img src="images/go.gif"
> width=162 height=37 alt="" border="0"></a>
> ```
>
> This is an effective design-time solution that causes the video signal to appear in the TV object. However, for the final production copy of content, the viewset.html page is not needed, because a live interactive TV link calls the page in TV mode.

To set the view property for the link to base.html:

1. Use the remote control to change the channel to another channel. Next, return to the previous channel. This will enable you to send a new link with the same URL values. This quirk will be explained in the "Where's the Icon?" section that appears later in this chapter.

2. In the Interactive TV Link Creator, fill in the Link URL and Display Name as described above in the instructions entitled "To create an interactive TV link to base.html on your local server." Remember to substitute the appropriate value for servername.

3. To the right of View As, click TV. You'll notice that when you change the View As property to TV, the Generated Link value at the bottom of the Interactive TV Link Creator changes to the following:

   ```
   <http://servername/humins/
   base.html>[v:t][n:Humongous Insurance][b6ed]
   ```

4. Notice that the abbreviated view attribute is set to [v:t] for TV.

5. Now click Trigger in Browser.

6. Click the Interactivity icon in the upper right corner of the screen. You will see the prompt shown in Figure 16.14.

Figure 16.14 *Prompt that appears when the view attribute is set to "TV."*

Notice that when the view attribute of the link is set to "TV," the prompt contains an Enhance TV button rather than a Go to Web Page button. This indicates that the page consists of interactive content that is meant to be shown in TV mode, as indicated in Figure 16.15.

NOTE The Microsoft TV equivalent to the Enhance TV button is a Go Interactive button.

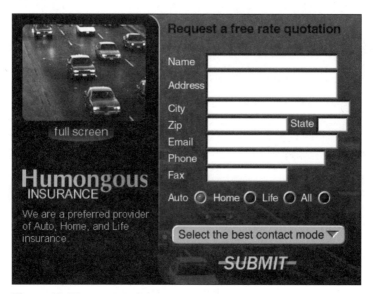

Figure 16.15 *Base.html shown in TV mode.*

There are two important points of interest here. First, the video signal appears in the upper left corner of the window. This is the TV object that is coded by the developer into the page. Second, the television window that previously appeared on the page has disappeared. This occurs when you specify TV, rather than Web, for the view attribute.

Where's the Icon?

In some cases, you may notice that the Interactivity icon fails to appear when you click the Trigger in Browser button in the Interactive TV Link Creator. The problem is that you are sending two links with the same URL. The solution, as above, is to change channels. However, the reason the icon fails to appear deserves attention. Here is what the specification adopted by the ATVEF has to say on the subject:

"When a broadcast data trigger is encountered, its URL is compared to the URL of the current page. If the URLs match and the trigger contains a script, the script should be executed. If the URLs match but there is no script, the trigger is considered a re-transmission of the current page and should be ignored. If the URLs do not match and the trigger contains a name, the trigger is considered a new enhancement and may be offered to the viewer. If the URLs do not match and there is no name, the trigger should be ignored."

NOTE   The term "broadcast trigger" is used by the ATVEF to mean an interactive TV link.

The bottom line is that if visible interactive TV links are sent repeatedly during a show, they are all ignored unless the viewer is new or has changed channels and returned. This implementation is more advantageous than one might assume, because if the TV receiver responded to multiple visible interactive TV links with the same URL, the Go Interactive prompt in Microsoft TV could pop up every few seconds, annoying the TV viewer.

THE LINK TYPE AND VIEW ATTRIBUTES

The Link Type attribute in the Interactive TV Link Creator window, as shown in Figure 16.16, was intended to give information about who the interactive TV link is related to.

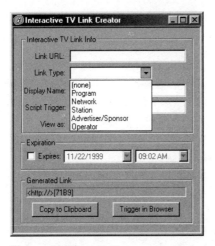

Figure 16.16 *Link Type attribute choices in WebTV Viewer.*

According to the specification adopted by the ATVEF, type is not one of the core attributes for a Transport A trigger. However, when the WebTV Viewer was created, trigger attributes were not standardized, so WebTV used the type property as follows:

- **Program**—the current television program
- **Network**—the broadcast network
- **Station**—the local affiliate station
- **Sponsor**—person or company paying for the programming
- **Operator**—the cable or satellite company

The problem with the WebTV implementation is that according to the specification adopted by the ATVEF, the type attribute may take on another meaning altogether. Following is an excerpt from the specification:

"In addition to the other attributes used in triggers…, ATVEF transport type A triggers must contain an additional attribute, "tve:". The "tve:" attribute indicates to the receiver that the content described in the trigger is conformant to the ATVEF content specification level. For example, [tve:1.0]. The "tve:" attribute can be abbreviated as the single letter "v". The version number can be abbreviated to a single digit when the version ends in ".0" (e.g. [v:1] is the same as [tve:1.0]). The "tve:" attribute is equivalent to the use of "type:tve" and "tve-level:" in SAP/SDP announcements in the transport type B IP multicast binding. This attribute is ignored if present in a trigger in transport B since these values are set in transport type B in the announcement. If the "tve:" attribute is not present in a transport type A trigger, the content described in the trigger is not considered to be ATVEF content."

This excerpt raises two apparent problems with creating links in the WebTV Viewer:

- First, according to the specification adopted by the ATVEF, Transport type A triggers must contain an additional attribute, "tve:" The problem is that there is no way to add the tve: attribute using the WebTV Viewer. Moreover, you need the WebTV Viewer, because it's a handy tool for calcuating the checksum value.

- Second, the tve: attribute can be abbreviated as the single letter "v." However, "v" also serves as an abbreviation for view (TV or Web) in the Interactive TV Link Creator. The letter "v" cannot simultaneously set view and tve: values. Also, the view attribute uses a letter value ("t" or "w"), whereas the tve: attribute uses an integer (1 for version 1.0 of ATVEF).

At first glance, this may seem like a problem. But thanks to the engineers at WebTV, it isn't. For example, the following link works on the WebTV Plus Receiver:

```
<http://servername/humins/base.html>[v:t][n:Humongous Insurance][b6ed]
```

However, the [v:t] attribute will be meaningless to ATVEF receivers other than WebTV Plus, so you will need syntax like the following:

```
<http://servername/humins/base.html>[v:1.0][n:Humongous Insurance][ecdc]
```

Fortunately, the WebTV Plus Receiver reacts the same to [v:t] or [v:1.0]. For both syntaxes, it pops up a button that says Go Interactive and then opens in TV mode.

The fact that the WebTV Plus Receiver handles both the [v:t] and [v:1.0] attributes in the same way is great news, but it doesn't resolve the issue of how to create links with the ATVEF [v:1.0] attribute. Simply substituting [v:t] for [v:1.0] doesn't work, because you need to recalculate the checksum value based on the changes. If your entire target audience consists of customers with WebTV Plus Receivers, this isn't a problem. However, if you want to reach a wider audience, you will need a new Interactive TV Link Creator tool. Alternatively, you can simply ship your content to an encoding vendor to encode the link for you.

This conflict between the ATVEF-required tve: attribute and the WebTV Plus–required view attribute is one of the main reasons for closely monitoring *http://www.microsoft.com/tv*. The new Microsoft TV Viewer tool should be out by the time this book is published, and most likely, it will allow you to create ATVEF-compliant links with checksum.

TRIGGER EXPIRATION

The expiration attribute in the Interactive TV Link Creator is useful for visible interactive TV links that are encoded onto tape. Taped versions of a TV show may float around indefinitely. But in many cases, you will only want to maintain the interactive TV data referenced by the links for a specific period of time. By specifying an expiration date, you instruct TV receivers not to show the interactive TV link prompt after the expiration date.

CREATING INTERACTIVE TV LINKS WITH SCRIPT TRIGGERS

Interactive TV links can be one of two kinds of triggers: a script trigger or a link trigger. In general, script triggers are invisible to the TV viewer, while link triggers pop up a prompt to let the TV viewer know that interactive TV content is available. As discussed earlier in this chapter, the way to make an interactive TV link visible (or to create what we call a link trigger) is to specify a name value for the link, which is defined in the Interactive TV Link Creator as the Display Name attribute.

To create a script trigger, you specify a script value for the interactive TV link, which is represented in the Interactive TV Link Creator as the Script Trigger attribute. In addition, you leave the name value, represented as Display Name, empty. When you define a Script Trigger value, you specify the name of a JavaScript function that resides on the URL you are referencing in the interactive TV link. Script triggers can be used to dynamically swap out pages and to dynamically change information on the current Web page. The following example shows how to build an interactive TV link that serves as a script trigger.

Creating a Script Trigger for Humongous Insurance

This section uses the Humongous Insurance demo to show how script triggers work. For this example, the script trigger will call a JavaScript function named out( ). The out( ) function loads a new page into the content frame of the Humongous Insurance demo. The out( ) function is placed on the base.html page, the page that defines the frameset for the Humongous Insurance content. The following code shows the out( ) function and the frameset defined on base.html:

```
function out()
{
    window.frames['content'].location.href = "testscript.html"
}

<frameset  cols="222,*" framespacing=0 frameborder=0 hspace=0 vspace=0
marginwidth=0 marginheight=0>
<frame name="tv" src="tvlogo.html" marginwidth="0"
marginheight="0" scrolling="no" noresize frameborder="no">
<frame name="content" src="HumForm.html" marginwidth="0"
marginheight="0" scrolling="no" noresize frameborder="no">
</frameset>
```

> **NOTE** Always place code that will be called by a script trigger in the top-level page of your interactive TV applications.

Now let's create an interactive TV link to load the base.html page for Humongous Insurance. After the base page is loaded with a visible interactive TV link, you send a interactive TV link with a script trigger.

To create an interactive TV link to base.html on your local server:

1. In the WebTV Viewer, open the Interactive TV Link Creator by clicking Show Link Creator Window from the TV menu.

2. In the Link URL box of the Interactive TV Link Creator, enter the URL ***http://servername/humins/base.html***, where servername is the name of your local server.

3. In the Display Name box, type **Humongous Insurance**.

4. To the right of View As, click TV.

5. Click the Trigger in Browser button, then click the icon, and then click the Enhance TV button to load the Humongous Insurance content.

Now that the base.html page is loaded, let's create a script trigger to fire the out( ) function. The Interactive TV Link Creator maintains the values previously specified, so you only need to change a few values.

To create an interactive TV link with a script trigger:

1. In the Link URL box of the Interactive TV Link Creator dialog box, leave the *http://servername/humins/base.html* value.

2. In the Display Name box, delete the Humongous Insurance text.

3. In the Script Trigger box, type **out( )**.

4. The View As value may be TV or Web. Just leave it TV for this example.

5. Click the Trigger in Browser button.

After you implement these steps, the testscript.html page will appear in the content frame as seen in Figure 16.17.

Figure 16.17 *Testscript.html in WebTV Viewer.*

Keep in mind that this script trigger example is simple but does demonstrate how you can call functions on a resident Web page with script triggers. By swapping more relevant pages than testscript.html into the content frame, there is no limit to how you can drive the interactive TV user experience.

Ending an Interactive TV Program with a Script Trigger

When the television program that is hosting your ITV content is over, you should clear the decks for any follow-on programming. To exit your application and put the viewer into full-screen video mode, do the following:

1. Place the gofull() function shown below in the top-level page of your application. In the Humongous Insurance example, base.html is the top-level page.

2. Create and run a script trigger using directions given earlier. Remember to substitute gofull( ) for out( ) in the Script text box in the Interactive TV Link Creator.

```
<http://servername/humins_rel/base.html>[v:t][s:gofull()][481e]

/* called by script trigger to go to full
   screen video
*/

function gofull()
{
    window.top.location.href = "tv:"
}
```

Triggering the gofull( ) function puts the TV receiver into full-screen TV mode, ready for the next interactive TV link to come down line 21 of the VBI.

Trigger Matching

The preceding example makes sending and executing a script trigger look a bit easier than it actually is. In a multipage application like Humongous Insurance, or any other application in this book, you can't be certain what page the user will be on when you send a script trigger. To help solve this problem, Microsoft TV provides a special syntax called trigger matching that enables you to send a script trigger to a series of pages, rather than an individual page. With trigger matching, for instance, the out( ) function used in the previous example could be placed on the humform.html page, rather than the base.html page. That way, the function will only fire when the humform.html page is loaded. For the TV receiver to execute the script on successive pages, it must determine that a match exists. The client determines that a match exists if the following conditions are met:

- The final character in the trigger is an asterisk (*), and
- The URL of the top-level page of the current interactive TV program matches the position of the asterisk.

    NOTE A complete domain must appear prior to the asterisk.

    For instance, if the trigger

```
<http://abc.com/mnf/*>[s:doscript(3)][2343]
```

was received by a Microsoft TV–based receiver that currently has one of the following pages loaded, the script will execute:

```
http://abc.com/mnf/score.html or
http://abc.com/mnf/stats.html
```

However, the script will not execute if the following page is loaded on the TV receiver:

```
http://abc.com/news/episode.html
```

This is because the root path abc.com/news does not match the string abc.com/mnf up to the "*" in the URL passed with the trigger.

INTERACTIVE TV LINK SEQUENCER

To enhance a show with interactive TV content, you can use a variety of interactive TV links. For example, some links may be directed to viewers who tune in late to a show. Other links might contain script triggers to dynamically swap out a page or to change the content of a page. The WebTV Viewer provides an Interactive TV Link Sequencer that you can use to practice the timing and execution of interactive TV links that you create.

To open the Interactive TV Link Sequencer:

In the WebTV Viewer, click Show Link Sequencer from the TV menu. The Interactive TV Link Sequencer appears, as shown in Figure 16.18.

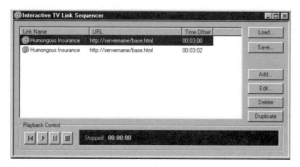

Figure 16.18 *Interactive TV Link Sequencer.*

ENCODING LINKS INTO THE VBI

Up to this point, you have been simulating interactive TV links. Now it's time to learn how to actually encode them. Links and triggers may be encoded directly into a video stream or onto tape. Unlike other forms of interactive TV transport, interactive TV links do not require an expensive high-resolution tape machine such as High-8. In fact, a VHS tape deck will do quite nicely for recording interactive TV links.

To help you encode interactive TV links, the companion CD provides a small link utility program that lets you encode interactive TV links with a Norpak encoder. The utility program files are located in the Resources\Utility folder of the companion CD. The link utility is designed to work with the Norpak Tess 3 and Tess 5 encoders. If you skipped chapters 2 and 3, you may wish to read them now to learn about the role of encoders in interactive TV. Chapter 15 presents information about Norpak and other interactive TV vendors.

With the link utility provided on the companion CD, you can paste interactive TV links from the WebTV Viewer into the link utility window, and then send the interactive TV link text strings to a Norpak encoder, which will encode them into the video stream.

To install the link utility:

1. On the companion CD, click Resources in the left-hand navigation panel, and then click the Interative TV Link Utility.

2. Double-click the Readme.txt file and read it carefully.

3. Close the window with the Readme.txt file.

4. Double-click the Setup.exe file and follow the on-screen prompts.

   NOTE You must connect the computer running the link utility to a Norpak encoder via serial cable and set the COM port baud rate correctly.

To encode an interactive TV link with the link utility:

1. In the WebTV Viewer, create an interactive TV link with the Interactive TV Link Creator.

2. Copy the link to the Windows clipboard by clicking Copy to Clipboard in the Interactive TV Link Creator.

3. Start the link utility program and then paste a link from the clipboard into the link utility window, as shown in Figure 16.19.

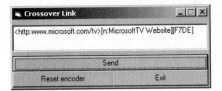

Figure 16.19 *Crossover Link player ready to encode.*

4. Click Send in the link utility window to send the interactive TV link to the encoder.

TRIGGER SYNTAX

So far, we have glossed over the details of trigger attributes in interactive TV links, mainly because the Interactive TV Link Creator handles most of the details for you. However, you may have occasion to use some of the trigger syntax described later in this chapter.

Following is an excerpt from the ATVEF Specification for Interactive Television 1.1:

"Triggers are real-time events delivered for the enhanced TV program. Receiver implementations will set their own policy for allowing users to turn on or off enhanced TV content, and can use trigger arrival as a signal to notify users of enhanced content availability. Triggers always include a URL, and may optionally also include a human-readable name, an expiration date, and a script. Receiver implementers are free to decide how to turn on enhancements and how to enable the user to choose among enhancements. Triggers that include a 'name' attribute may be used to initiate an enhancement either automatically, or with user confirmation. The initial top-level page for that enhancement is indicated by the URL in that trigger. Triggers that do not include a 'name' attribute are not intended to initiate an enhancement, but should only be processed as events which affect (through the 'script' attribute) enhancements that are currently active. If the URL matches the current top-level page, and the expiration has not been reached, the script is executed on that page through the trigger receiver object...When testing for a match, parameters and fragment identifiers (i.e. characters in the URL including and following the first '?' or '#' character) in an URL are ignored."

Triggers are text based, and their syntax follows the basic format of the EIA-746A standard (7-bit ASCII, the high-order bit of the first byte must be "0"). Note that triggers follow the syntax of EIA-746A, but they may be transported in multicast IP packets or other transport, rather than via the EIA-608 system.

All triggers defined in the specification adopted by the ATVEF are text based and must begin with ASCII '<'. All other values for the first byte are reserved. These reserved values may be used in the future to signal additional messages that are not text based. Receivers should ignore any trigger that does not begin with the '<' in the first byte.

The general format for triggers (consistent with EIA-746A) is a required URL followed by zero or more attribute/value pairs and an optional checksum:

```
<url> [attr₁: val₁][attr₂:val₂]...[attrₙ:valₙ][checksum]
```

- All characters are based on the ISO 8859-1 character set (also known as Latin-1 and compatible with US-ASCII) in the range 0x20 and 0x7e. Any need for characters outside of this range (or excluded by the attribute limits shown below) must be encoded using the standard Internet URL mechanism of the percent character (%) followed by the two-digit hexadecimal value of the character in ISO 8859-1.

- The trigger begins with a required URL:

  ❑ **<url>** The URL is enclosed in angle brackets: <http://www.microsoft.com/tv>. Although any URL can be sent in this syntax, ATVEF content level 1 only requires support for http: and lid: URL schemes. Triggers begin with the URL inside of arrow tags, <url>. The URL must be completely resolved. Completely resolved means all there, including the .com and the same path you would use in a Web browser.

- The following attribute/value pairs are defined:

  ❑ **[name:***string***]** The *name* attribute provides a readable text description (e.g.[name:Find Out More]). The string is any string of characters between 0x20 and 0x7e except square brackets (0x5b and 0x5d) and angle brackets (0x3c and 0x3e). The *name* attribute can be abbreviated as the single letter "n" (e.g.[n:Find Out More]).

  ❑ **[expires:***time***]** The *expires* attribute provides an expiration date after which the link is no longer valid (e.g.[expires:19971223]). The time conforms to the ISO 8601 standard, except that it is assumed to be UTC unless the time zone is specified. A recommended usage is the form *yyyymmddThhmmss*, where the capital letter "T" separates the date from the time. It is possible to shorten the time string by reducing the resolution. For example *yyyymmddThhmm* (no seconds specified) is valid, as is simply *yyyymmdd* (no time specified at all). When no time is specified, expiration is at the beginning of the specified day. The *expires* attribute can be abbreviated as the single letter "e" (e.g.[e:19971223]).

❑   [**script**:*string*]   The *script* attribute provides a script fragment to execute within the context of the page containing the trigger receiver object (e.g.[script:shownews()]). The string is an ECMAScript fragment. The *script* attribute can be abbreviated as the single letter "s" (e.g.[s:shownews()]). An example of a *script* attribute used to navigate a frame within a page to a new URL: [script:frame1.src="http://atv.com/f1"]

■   The optional checksum must come at the end of the trigger. (Note: EIA-746A requires the inclusion of a checksum to ensure data integrity over line 21 bindings. In other bindings, such as IP, this may not be necessary and is not required.)

❑   [**checksum**]   The checksum is provided to detect data corruption and is required for interactive TV links.

The checksum is provided to detect data corruption. To compute the checksum, adjacent characters in the string (starting with the left angle bracket) are paired to form 16-bit integers; if there are an odd number of characters, the final character is paired with a byte of zeros. The checksum is computed so that the one's complement of all of these 16-bit integers plus the checksum equals the 16-bit integer with all 1 bits (0 in one's complement arithmetic). This checksum is identical to that used in the IP (described in RFC 791); further details on the computation of this checksum are given in IETF RFC 1071. This 16-bit checksum is transmitted as four hexadecimal digits in square brackets following the right square bracket of the final attribute/value pair (or following the right angle bracket if there are no attribute/value pairs). The checksum is sent in network byte order, with the most significant byte sent first. Because the checksum characters themselves (including the surrounding square brackets) are not included in the calculation of the checksum, they must be stripped from the string by the receiver before the checksum is recalculated there. Characters outside the range 0x20 to 0x7e (including the second byte of two-byte control codes) are not included in the checksum calculation.

Other attributes may be defined at a later date. However, all other single character attribute names are reserved. Receivers should ignore attributes they do not understand.

Using the description above, all the following are valid trigger strings:

```
<http://xyz.com/fun.html>
```

```
<http://xyz.com/fun.html>[name:Find out More!]
```

```
<lid://xyz.com/fun.html>[n:Find out More!]
```

```
<lid://xyz.com/fun.html>[n:Fun!][e:19991231T115959] [s:frame1.src="http://
atv.com/frame1"]
```

```
<http://www.newmfr.com>[name:New][c015]
```

> **NOTE**   If a trigger does not contain a [name:] attribute, the enhancement referenced by the trigger should not be presented to the user. Also, triggers follow the syntax of EIA-746A, but they may be transported in multicast IP packets or other transport alternatives to the EIA-608 system.

WHAT'S NEXT

In this chapter, we demonstrated how to use the Interactive TV Link Creator in the WebTV 2.0 Viewer to create interactive TV links. In addition, we described how to create two kinds of interactive TV links: those with link triggers and those with script triggers. Finally, we explained how to use a link utility program to insert and test links at design time. In the next chapter, we will examine how the Humongous Insurance content covered in this chapter is created. Perhaps more importantly, we will explain how to implement e-commerce applications for Microsoft TV.

Part IV

Microsoft TV
E-Commerce

```
var oldsrc = ""

function featureload()
{
        parent.frames["content"].location.href="exotic_feature1.html";
        document.btn_feature.src = featuredon.src;
        if (oldimage != document.btn_feature)
        {
                oldimage.src = oldsrc
        }
        oldimage = document.btn_feature
        oldsrc = featuredoff.src
}

function reservationsload()
{
        parent.frames["content"].location.href="exotic_reservations1.html";
        document.btn_reservations.src = reservationson.src
        if (oldimage != document.btn_reservations)
        {
                oldimage.src = oldsrc
        }
        oldimage = document.btn_reservations
        oldsrc = reservationsoff.src
}
```

Chapter 17

Creating Forms for Microsoft TV Content

In This Chapter

- Where to Find Sample Content for This Chapter
- Overview of the Interactive Content in This Chapter
- Creating the Humform.html Form Page
- Scripting to Form Elements
- Creating the Thanks.asp Page
- Using ASP with HTML Tags

A key feature of Microsoft TV is the interaction that occurs between viewers and content. The HTML form is the prime vehicle by which this interaction takes place. Viewers use an HTML-based form to enter information about themselves or their preferences into input elements, such as text boxes, radio buttons, option groups, and text areas. On set-top boxes, viewers enter information via a wireless keyboard or, for the truly persistent, a virtual keyboard that pops up on the TV screen much like a calculator on a computer screen and is manipulated by an infrared remote control.

After entering information, a viewer typically activates a button that fires a *submit* event. Information is then passed to the server, where it is processed, and a return page is assembled.

Just about every HTML book published has covered the topic of HTML forms and form elements. It would be fair to say that if your office develops commercial Web pages you have plenty of tomes chock full of details on HTML forms, so we will not recover that ground here. In this chapter, we take a look at how you create the forms and the form elements for the mouse-less world of interactive TV. We will emphasize a set of special form input attributes for the Microsoft TV 1.0 browser that are key to the development of professional-looking interactive TV e-commerce application forms. We will also provide instructions for building the Active Server Page (ASP) that returns a message to the viewer.

WHERE TO FIND SAMPLE CONTENT FOR THIS CHAPTER

To get the most out of the material in this chapter, we recommend that you view the sample files on the *Building Interactive Entertainment and E-Commerce Content for Microsoft TV* companion CD and Web site.

- To view the Humongous Insurance content on a set-top box running Microsoft TV or on the Web, go to *http://www.microsoft.com/tv/itvsamples*.

- To view html files associated with this chapter, see the "Templates" section of the companion CD.

- For source files for the Humongous Insurance application, see the "Humongous Insurance" topic of the "Microsoft TV E-Commerce" section of the companion CD.

OVERVIEW OF THE INTERACTIVE CONTENT IN THIS CHAPTER

This chapter describes the workings of a fictitious interactive advertisement for Humongous Insurance. The form elements of the advertisement are shown in the right frame of Figure 17.1. The rest of the files that make the ad work are listed below. They were constructed using dimensions and techniques covered in earlier chapters, so their roles will be covered only briefly. The file descriptions are provided to reveal how individual files interact to form a complete interactive TV application. Feel free to use or change the files in this demo for your own purposes.

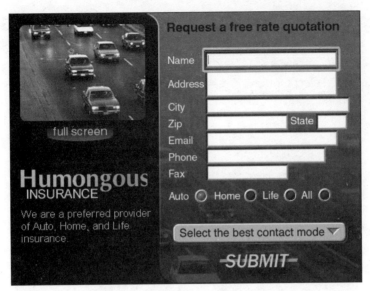

Figure 17.1 *Pages used in the Humongous Insurance ad.*

The Humongous Insurance demo consists of the following Web pages:

- **base.html**—defines the frameset and the frames for the Humongous Insurance demo.

- **tvlogo.html**—defines the Web page in the left frame. This page contains the TV control, the logo, and advertising. The tvlogo.html page is always visible in the frame.

- **humform.html**—appears in the right frame. It contains the form and form elements discussed in this chapter.

- **thanks.asp**—a page dynamically created by ASP technology in response to the *submit* event on humform.html. Thanks.asp renders information contained in the form elements and lets the viewer know that desired behavior is occurring. You will hear more about thanks.asp later in this chapter.

- **cssPC.css and cssTV.css**—cascading style sheets (CSS) used to render proper fonts to the platform, as described in Chapter 23, "CSS Level 1 Support for Microsoft TV."

- **viewset.html**—a test page that acts as a proxy for the broadcast link that will be sent along with a video signal as part of the vertical blanking interval (VBI). The VBI and other issues surrounding how content reaches the TV client are covered in Chapter 15, "Fundamentals of Delivering Interactive TV Content," and Chapter 16, "Creating Interactive TV Links."

CREATING THE HUMFORM.HTML FORM PAGE

The humform.html page defines the form and form elements used for the Humongous Insurance demo. Figure 17.2 shows this page without the frameset that contains it. Notice how the Select Option Group and Radio Button form elements render slightly differently on Microsoft TV from what you may be used to on a personal computer. This variation in the way controls are rendered is built into the two platforms. You do not need to do any special coding to make the Select Option Group show up in the Microsoft TV client, as seen in Figure 17.1.

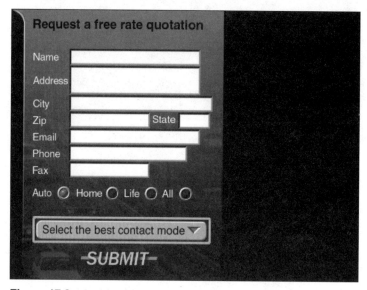

Figure 17.2 *The humform.html page on a personal computer.*

How Humform.html Was Created

The humform.html page is easy to create, as shown by the following sample code. Nevertheless, there are some special attributes and techniques used for this page that deserve special attention. We will provide details about these techniques in the following sections.

```
<html>
<head>

<link rel="stylesheet" type="text/css" href="cssTV.css">
<script>

/* code to load the proper style sheet for the platform encountered */

var bName = navigator.appName
if (bName == "Microsoft Internet Explorer")
```

```
{
   document.write("<Link Rel='stylesheet' type='text/css' ↴
      href='cssPC.css'>");
}

/* sendForm() function is called by the click event of the Submit
image button */

function sendForm()
{
   thisform = document.forms[0]
   thisform.submit()
}
</script>

<title>Humongous Insurance Form </title>
</head>

<body bgcolor="#003366">
<div style="position:absolute; top:16; left:16; width:306; height:388;
background-color:#828ECF;">

<div style="position:absolute;top:12;left:15;" class="clsTitle">

Request a free rate quotation

</div>

<div style="position:absolute;top:42;left:5;>

<form name="frmGetQuote" method=post action="thanks.asp">

<!-- table holding text box inputs  -->
<table border=0  cellpadding="0" cellspacing="0">
<tr><td class="clsDescription"> Name </td>
<td> <input type="text" size=20 name="txtLastName"></td></tr>

<tr><td class="clsDescription">Address  </td>
<td> <textarea rows=2 cols=20 name="txtAddress"> </textarea></td></tr>

<tr><td class="clsDescription">City  </td>
<td><input type="text" size=22 name="txtCity" ></td></tr>
<tr><td class="clsDescription"> Zip </td>
<td><input type="text" size=12 name="txtZip">
<font class="clsDescription"> State </font>
<input type="text" size=4 name="txtState"></td></tr>
```

(continued)

```
<tr><td class="clsDescription"> Email</td>
<td> <input type="text" size=20 name="txtEmail"></td></tr>
<tr><td class="clsDescription">Phone </td>
<td><input type="text" size=12 name="txtPhone"></td></tr>
<tr><td class="clsDescription">Fax </td>
<td><input type="text" size=12 name="txtFax"></td></tr>
<tr><td height="10"></td></tr>

</table>

<!-- table holding radio button group  -->
<table align="center">
<tr><td class="clsDescription">Auto </td>
<td><input type="radio" name="radType" value="auto" checked> </td>
<td class="clsDescription">Home</td>
<td> <input type="radio" name="radType" value="home"> </td>
<td class="clsDescription">Life</td>
<td><input type="radio" name="radType" value="life"> </td>
<td class="clsDescription">All</td>
<td><input type="radio" name="radType" value="all"> </td>
</tr>

</table>

<table border=0  cellpadding="0" cellspacing="0" align="center">
<tr><td height="10"></td></tr>
<tr><td><select name="selContactMode" size="1" autoactivate >
<option value="1" selected> Select the best contact mode </option>
<option value="e-mail"> E-mail me a quote </option>
<option value="phone"> Phone my quote to me </option>
<option value="fax"> Fax me my quote </option>
<option value="s-mail"> Snail mail me a quote </option>
</select></td></tr>
<!-- Simple 10 pixel spacer -->
<tr><td height="10"></td></tr>

<!-- Submit image button -->
<tr><td align="center">
<a href="JavaScript:onclick=sendForm()">
<img src="images\submit.gif" border="0">
</a></td>
</tr>
</table>

</form>

</div>
</div>

</body>
</html>
```

The Post vs. the Get Method

The first thing you need to decide when using forms to gather information is how to set the method attribute of the opening form tag.

```
<form name="frmGetQuote" method=post action="thanks.asp">
```

Setting the form *method* attribute to *post* means information is sent to the server as part of the HTTP header in distinct chunks. The alternative to post is *get*, which can serve as a value for the method of a form. The get method sends information as part of a long delimited string that is contained within the URL. Common Gateway Interface (CGI) programs have traditionally used the get method to pass information from the client Web page to the server. ASPs can accommodate either post or get values. Accordingly, if you submit a search at a Microsoft home page, you might see something like the following in the browser URL box, or you might see nothing at all.

```
http://search.microsoft.com/us/
results.asp?nq=TRUE&SName=&SPath=&SCatalog=&qu=DTV&Boolean=PHRASE&Finish=
Search+Now%21&intCat=10
```

Truncating values onto the end of a URL has disadvantages in e-commerce situations. The field values sent via the get mode are clearly visible and can be easily intercepted while in transit. Moreover, there is a 1,000-character limit to the amount of information that may be sent via get.

The pages in this demo will use the post method to submit data fields and the ASP Request.Form collection to interpret them. This approach has two key advantages: it keeps form data hidden, and it is easier to use than the get method.

The Form Action Attribute

The other vital bit of information every good opening <form> tag must contain is a value for the *action* attribute.

```
<form name="frmGetQuote" method=post action="thanks.asp">
```

The action attribute of a form tells the server what page to return after it processes values submitted by form input elements. In this example the server is instructed to return a page named thanks.asp. You will see how ASP can be used to dynamically create a responsive "Thank You" page later in this chapter.

The page humform.html houses a form whose action attribute calls another page: thanks.asp. Values for the action attribute of a form can and often do name the same page that houses the form. For example, the action attribute of the form on humform.html could be humform.html. In this circumstance, when the form submits information to the server, the server does something with it, such as writing the information to a database. Next, the server returns the same page, usually with some confirmation of desired behavior or a request for more information.

Form Input Elements and Attributes

Most of the labels near input elements on the Humongous Insurance form have styles set by the <td> cell they live in.

```
<td class="clsDescription">All</td>
<td><input type="radio" name="radType" value="all"> </td>
```

The radio button input with the label "All" is controlled by a CSS class named clsDescription. This class is part of the linked style sheet named cssPC.css or cssTV.css, depending on what browser is encountered.

NOTE   See Chapter 23, "CSS Level 1 Support for Microsoft TV," for a complete discussion of using CSS style sheets in interactive TV coding.

Style classes may also be set in <font> tag sets. For example, because the input labels "Zip" and "State" reside in different cells, the "State" cell must have its class set separately.

```
<tr><td class="clsDescription"> Zip </td>
<td> <input type="text" size=12 name="txtZip">
<font class="clsDescription"> State </font>
<input type="text" size=4 name="txtState" ></td></tr>
```

Every cell must be treated separately within an HTML table, when it comes to CSS style classes in Microsoft TV.

Input Text Boxes

Most of the inputs used on the Humongous Insurance forms are simple text boxes like the one shown below.

```
<input type="text" size=22 name="txtCity">
```

The Name Attribute

The above text type input named txtCity enables viewers to identify the city they live in. It is essential to set the name attribute of each form input element. The value of this attribute is used in code to process input from the form. It is a good idea to use descriptive names with three-letter prefixes, such as txtCity for the City field, when naming your input fields. The three-letter prefix should give a clue to the expected data type that the fields will hold. We will return to this topic later in this chapter in discussing the return of a customized "Thank You" page to viewers. As you will see in Chapter 20, "Creating ASP-ADO Code to Interface with the Database," data types must match when writing to a back-end database. In Chapter 20, you will also see much more on how to ensure data type integrity through inline validation code.

Size Attribute

Text boxes are used for entering single rows of data. The width of a text box is controlled by its *size* attribute. The text box named txtCity is 22 characters wide. Keep the design-space limitations imposed by the TV space in mind when you set width values for form elements.

Class and Usestyle Attributes

The Microsoft TV client browser only picks up CSS class attributes from the <div> tags surrounding elements or from a special Microsoft TV–only attribute named *usestyle*. To affect the cursor in a form element on Microsoft TV, you may surround it with a <font> tag set or use the usestyle attribute. The usestyle attribute will render text in a form element, using the style currently in effect on the page.

```
<font color=blue>
<input type=text name=txtTest cursor=green usestyle>
</font>
```

Input TextArea Boxes

The *textarea* input provides the best way to handle multiline text in Microsoft TV forms. Take a quick look at the textarea input used in this form:

```
<textarea rows=2 cols=20 name="txtAddress">Four score and seven...
</textarea>
```

Text area controls, unlike text boxes, require an ending tag. If you want default text to appear in the text area, simply place it between the beginning and ending tags. For more information about this subject, see the section "Scripting to Form Elements" later in this chapter.

```
<textarea rows=2 cols=20 name="txtAddress">
Four score and seven chapters ago I was blissfully ignorant of ITV coding.
</textarea>
```

The size of a text area is controlled by its *cols* and *rows* attributes. The text area shown in the Humongous Insurance form is 2 rows high by 20 characters wide. In the personal computer–based Microsoft Internet Explorer clients, a scroll bar allows access to text area text that exceeds the set size of the control. In Microsoft TV, viewers select the control and then press the Edit key on their wireless keyboard to do the same thing. No scroll bars are used on the Microsoft TV implementation of text area controls.

Insetselection Attribute

The yellow box indicating which input has focus in Microsoft TV can sometimes obscure labels or other nearby design elements, as shown in Figure 17.3.

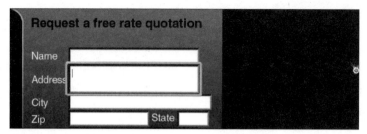

Figure 17.3 *The selection box can obscure design elements.*

The problem shown in Figure 17.3 may be alleviated by using another special Microsoft TV attribute, called *insetselection*. The insetselection attribute may be used with any element that may gain focus on the page. This includes <a> anchor tags as well as form elements. Adding the insetselection attribute to the opening tag of the text area control positions the yellow selection box on the inside edge of the control, as shown in Figure 17.4.

```
<textarea rows=2 cols=20 name="txtAddress" insetselection>
```

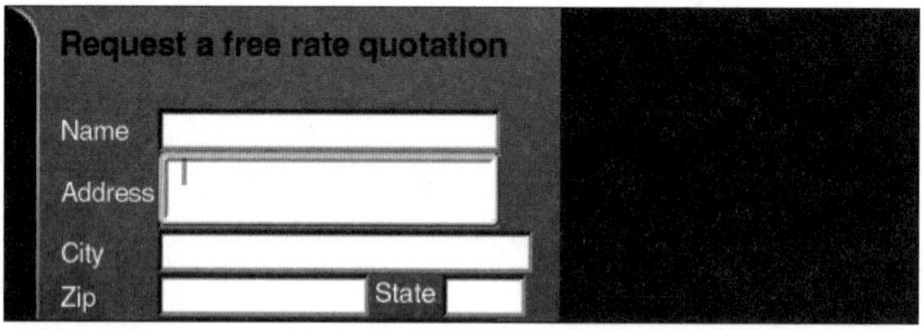

Figure 17.4 *Selection box set on the inner edge of the text area.*

Physical, Virtual, Nohardbreaks, and Nosoftbreaks Attributes

Another important attribute is *wrap*, which determines how text is handled and submitted in text area controls. In personal computer–based clients like Internet Explorer, setting the wrap attribute to *physical* or *virtual* causes text in the text area to create a new line or "wrap" when the width of the control is reached. The physical setting submits the text to the server with wrapping included, while virtual displays wrapped text but submits it as a solid string. The physical and virtual settings—as they apply to wrap—are not supported by Microsoft TV. In their place, Microsoft TV provides *nohardbreaks* and *nosoftbreaks*. The nohardbreaks attribute prevents the viewer from entering hard breaks in a text area. If the nosoftbreaks attribute is used, breaks are not submitted with the text to the server.

```
<textarea rows=2 cols=20 name="txtAddress" insetselection wrap="virtual" nosoftbreaks>
```

All personal computer–based clients, such as Internet Explorer, will ignore attributes like nosoftbreaks and insetselection without bad results.

The Nohighlight Attribute

Because input controls such as text boxes and text areas give the viewer a cursor, you may opt to dispense with the yellow selection box altogether. You can accomplish this by using Microsoft TV's *nohighlight* attribute. For example, the following code will render a text box in which viewers can enter their names without using the yellow selection box, as seen in Figure 17.5.

```
<input type="text" size=20 name="txtLastName" nohighlight>
```

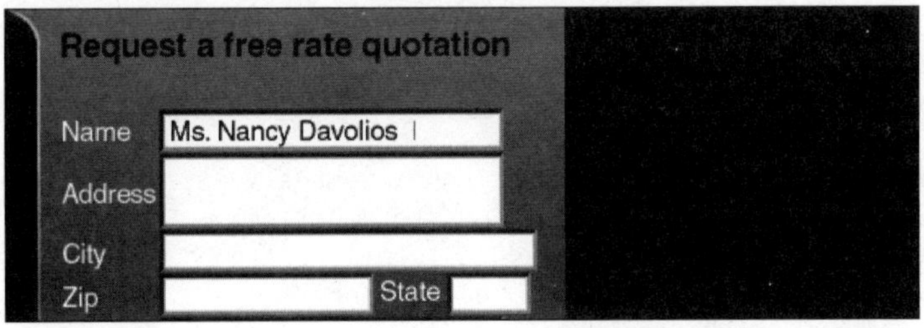

Figure 17.5 *Text input using the nohighlight attribute.*

As with all special Microsoft TV 1.0 attributes, nohighlight has no effect on personal computer–based clients.

Radio Buttons

Radio buttons work the same way on Microsoft TV as they do on personal computer platforms. The mere fact that you set all their name attributes to the same value means that only one choice is possible, which is what makes them radio buttons.

```
Auto<input type="radio" name="radType" value="auto" checked>
Home<input type="radio" name="radType" value="home">
Life<input type="radio" name="radType" value="life">
All<input type="radio" name="radType" value="all">
```

When the form is submitted, the value of the selected radio button is sent. Notice that the *checked* attribute of the radio button value designated as = "auto" for automobile is set. Only one button in the group may have its checked attribute set. The value of the checked button becomes the default value for the group. When the form is submitted, the value of the selected radio button is sent as the input from the radio group. No special coding is required to make radio buttons take on the TV persona shown in the illustrations.

Select Option Groups

Select Option Groups, shown in Figure 17.6, are used to provide lists of options to the viewer. They are a great way of providing a large number of choices within the limited confines of the 560-by-420 design space dictated by the TV monitor.

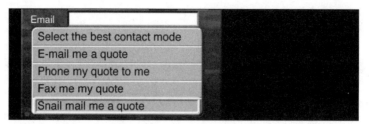

Figure 17.6 *Select Option Groups expanded in Microsoft TV.*

The <select></select> tag sets act as containers for option inputs within the group.

```
<select name="selContactMode" size="1">
<option value="1" > Select the best contact mode </option>
<option value="e-mail"> E-mail me a quote </option>
<option value="phone"> Phone my quote to me </option>
<option value="fax"> Fax me my quote </option>
<option value="s-mail"> Snail mail me a quote </option>
</select>
```

Size Attribute

The size attribute of the opening select tag determines how many of the options are visible without any viewer action.

Make Select Option Groups Obvious

Some users of Microsoft TV may be completely flummoxed by something as simple as a drop-down group of options, so it is important to label option groups as clearly as possible. One way to make it clear that the viewer has more than one choice is to set the *size* attribute higher than 1; the result is shown in Figure 17.7.

Figure 17.7 *Multiple options in view.*

```
<select name="selContactMode" size="3">
```

Showing more than one option makes it clear to viewers that more than one choice is available. If you like, you may wish to allow the viewer to select more than one item from the list. This can be accomplished by setting the *multiple* attribute, as shown below:

```
<select name="selContactMode" size="3" multiple>
```

Unfortunately, there are some disadvantages to showing more than one option. In particular, expanded Select Option Groups take up more vertical space than the single-line counterparts. This can cause elements to be pushed off the page. If you do decide to reveal more than one option at a time, there is a special behavior of Microsoft TV 1.0 to take into account.

The AutoActivate Attribute

Microsoft TV viewers do not have computer mouse devices. They select things by using the direction arrows on their infrared remote control keyboard or handheld remote. If a viewer navigates down to a Select Option Group with multiple options in view, the group as a whole will have focus, not the individual options within it. This behavior is illustrated in Figure 17.8.

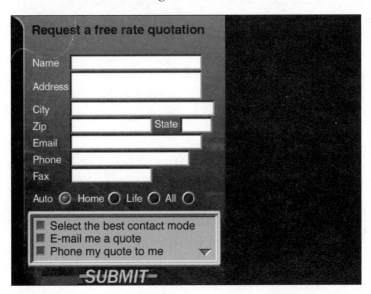

Figure 17.8 *The Select Option Group has focus, rather than one of the options.*

Imagine you are the viewer of the Humongous Insurance form shown in Figure 17.8. You just used your Down Arrow button to navigate to the Select Option Group, and you want to choose a response mode for your insurance quote. If you press the Tab key, you move to the Submit button with no choice made. If you press the Up Arrow or Down Arrow key, you similarly move to another control on the page without getting the opportunity to make a choice. So, how do you get into the Select Option Group? The answer is to press the Enter key or the Go button in the center of the handheld remote control. Next, you must use an arrow key to give focus to one of the options. Finally, you must press the Enter key again to select the option. Selecting an option gives focus back to the outer select group, so if you wish to change your option choice you must navigate this whole puzzle again. The process is extremely counterintuitive and might discourage even seasoned computer users.

Thankfully, the developers of Microsoft TV created a special attribute to the <select> tag that helps make multiline Select Option Groups intuitive to use. The attribute, *AutoActivate*, is shown below in code:

```
<select name="selContactMode" size="3" autoactivate>
```

Using the AutoActivate attribute with a Select Option Group makes the control behave in a more intuitive manner. Navigating to the Select Option Group automatically selects one of the options, as shown in Figure 17.9.

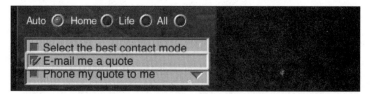

Figure 17.9 *Select Option Group with the AutoActivate attribute.*

Notice that the area around the Select Option Group did not get focus. One of the options in the group is "auto-activated" when a viewer navigates to the group.

Pressing the Enter key or the Go button selects the option. To change a selection, the viewer simply navigates to the new choice via an arrow key and presses the Return key again. This clears the old selection and selects the new one.

Nocheckboxes Attribute

If your design sensibilities reject the check marks rendered next to WebTV option inputs, you can use a special attribute, nocheckboxes, to remove them. Unfortunately, check marks are an all-or-nothing proposition. You cannot, for example, remove the check mark in front of "Select the best contact mode" while leaving other check marks in place.

The nocheckboxes attribute is shown below in code:

```
<select name="selContactMode" size="3" autoactivate nocheckboxes>
```

Figure 17.10 shows the result of using the nocheckboxes attribute—the check marks are gone.

Figure 17.10 *The nocheckboxes attribute.*

Next, we will introduce some concepts regarding scripting to form elements. This topic will be explored in greater detail in chapters 18, 19, and 20, which discuss the scripting used in the e-commerce solution named Bob's Pizza.

SCRIPTING TO FORM ELEMENTS

This section will cover scripting to several types of form elements. We will begin with scripting to the *submit* method of a form object to mimic the behavior of a submit-type input button. This technique is useful when you wish to use a graphic image rather than a button to submit form content to a server. We will also discuss how to prompt viewers for proper input, using as an example the detection of which item in a select list has been chosen by a viewer. Finally, we will explore the importance of using the textarea control for changing descriptive text on-screen.

Script Overview

Scripting to form elements involves three steps:

1. Setting a function name to an event recognized by the input element.

2. Creating a function.

3. Scripting the function to (1) identify the form; (2) specify the element within that form; and (3) use one of the element's properties or methods to do useful work.

Using the Submit Method

The Humongous Insurance form uses an image instead of a form-input Submit button, as shown in Figure 17.11. For this reason, it requires scripting in order to fire the submit event.

Figure 17.11 *The Submit button.*

```
<a href="JavaScript:onclick=sendForm()">
<img src="images\submit.gif" border="0">
</a>
```

Notice that the image must be wrapped in an <a> anchor tag so Microsoft TV will recognize it as a hot link. The *onClick* event is set in the anchor *href* attribute to fire a function named sendForm(). Let's examine how the sendForm() function is used to submit this form.

```
<script>
/* sendForm() function is called by the onclick event of the anchor
   around the submit image button */

function sendForm()
{
   thisform = document.forms[0];
   thisform.submit();
}
</script>
```

The Document Object Model (DOM) allows programmers to identify a form on the page using either the forms array ordinal number or the name of the form as set using the name attribute. The following two lines of code are equivalent:

```
thisform = document.forms[0];
thisform = document.frmGetQuote;
```

The sendForm() function begins by filling a variable, thisform, which represents a form on the page.

```
thisform = document.forms[0];
```

We have found the *forms[0]* array syntax to be the most reliable way to reference a form. Make sure you use the square [ ] brackets rather than the oval ( ) ones when scripting to arrays in Microsoft TV.

Once you have a handle on the form, you may use the built-in submit() function to submit the contents of the form to the server.

```
thisform.submit()
```

Scripting Cues for Viewers

You may find the DOM and scripting to form elements useful in giving viewers cues to properly fill out a form. The basic syntax for referring to any form element is:

```
Thisform.namedElement
```

For example, suppose a viewer has chosen to be contacted via e-mail, but has left the e-mail text input box empty. The following code could remind him to enter a value before submitting the form:

```
<select name="selContactMode" size="3"  onchange="chkField()">
<option value="e-mail"> E-mail me a quote </option>
<script>

function chkField()
{
   thisform = document.forms[0];
   thislist = thisform.selContactMode;
```

```
thisoption = thislist.selectedIndex;
thisword = thislist.options[thisoption].value;

if(thisword == 'e-mail')
{
   thisform.txtEmail.value = "Enter e-mail address";
}
}
</script>
```

As the above script shows, the variable thisform refers to the form on a page, using the same technique that we applied to the submit method. The variable thislist refers to the Select Option Group with its name attribute set to selContactMode. The selectedIndex property of the Select Option Group referred to by this list is used to fill the variable thisoption with a zero-based integer value representing the viewer's choice. To make the code more self-documenting, the value property of the *options* array set to the integer held by thisoption is polled to get a human readable value. For example, if the viewer chose "E-mail me a quote" from the Select Option Group, thisoption will be the integer "1." The value of the "1" integer in the *options* array is "e-mail," because that is the second option in the group (zero-based) and its value attribute is set to "e-mail," as shown in Figure 17.12.

```
<option value="e-mail"> E-mail me a quote </option>
```

The decision block is entered, and a prompt is written to the txtEmail text input field in the form.

```
thisform.txtEmail.value = "Enter e-mail address";
```

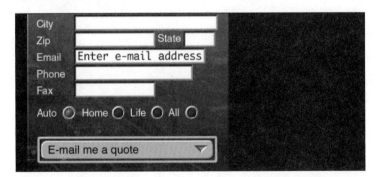

Figure 17.12 *Scripting to form elements can cue viewer input.*

If you were to fully implement the code solution shown for the chkField function, you would add If Then decision branches for the other options in the Select Option Group. Remember that the commonly used Switch and Select Case constructs are not part of the specification adopted by the Advanced Television Enhancement

Forum (ATVEF) for ECMA 262 Language (ECMAScript). To ensure that your code will work on all ATVEF receivers, use If Then rather than Switch or Select Case.

Scripting to the Textarea Control

Changing text in a textarea control in code is surprisingly important to e-commerce in interactive TV development, due to a restriction imposed by the ATVEF. The specification issued by the ATVEF asserts that receivers need only recognize Level 0 DOM and ECMAScript 1.1. The upshot for interactive TV developers is that TV client receivers might potentially be built that would not recognize the document.all.elementID method of referring to tags or divisions on your page. The *document.all* array is used to reference elements on the page that are not among the hard-coded arrays. Examples of arrays other than the hard-coded arrays include the *forms* and *images* arrays. Most HTML 4.0 developers use stacked <div> areas and the visibility attribute of the style object to hide and show appropriate text in response to page events. For example, the following code might be used to stack divisions in the same x-y coordinate and change which division the viewer sees in response to some event:

```
<script>
/* called by the onclick event of the body tag it reverses the visibility
   of these two divisions */

function change()
{
   document.all.divGreet.style.visibility = "hidden";
   document.all.divSell.style.visibility = "visible";
}

</script>
<body onclick="change()">
<!-- two divisions set on the same x-y spot, one visible and one hidden -->

<div id="divGreet" style="position:absolute; top:100;  left:100;
visibility:visible;">

Welcome to Humongous Insurance

</div>
```

```
<div id="divSell" style="position:absolute; top:100;  left:100;
visibility:hidden">

Don't just stand there, buy something

</div>

</body>
</html>
```

Unfortunately, this solution for dynamically changing text may not work on a TV client that is capable of recognizing only Level 0 DOM and ECMAScript 1.1.

NOTE Microsoft TV exceeds the minimum standards of the specification issued by the ATVEF and supports scripting to <div> elements through the *all* array.

To ensure that your pages will work with any TV client, you can script changing content to a textarea. We will illustrate this point using code from the Bob's Pizza example in Chapter 19, "Building Bob's Order Entry Page." This ad uses a text area to describe the available pizza toppings. The *onChange* event of a Select Option Group is used to write descriptions that match toppings selected by the viewer.

```
<textarea name="txadesc" readonly rows=3 cols=30> Sausage, olives, mushrooms,
peppers, tomatoes, garlic and pesto on a garlic crust
</textarea>
```

The opening value for the text area with the name attribute set to txadesc is shown in Figure 17.13.

Figure 17.13 *Opening text-area text in Bob's Pizza.*

When viewers change the entries in the pizza selection box, an *onChange* event fires and calls a function named calcprice(), as shown in Figure 17.14.

Figure 17.14 *The* onChange *event in the Select Option Group changes the textarea value.*

As the following coding shows, scripting to a text area provides a good ATVEF-compliant way to dynamically change text on a page.

```
if (thistype == 'kahuna')
{
    thispizza = "Canadian bacon, pineapple, mushrooms, and a ↴
        sprinkling" + " of macadamia nuts on a crispy natural crust";
    thisform.txadesc.value = thispizza;

    if(thissize == '12')
    {
        thisform.txtPrice.value = "$" + (13.33 * thismany)
        thisform.hidProdID.value = 10
    }
```

Chapters 18, 19, and 20 will provide a more detailed discussion of scripting to form elements for interactive TV solutions. For now, let's move on to the return page of Humongous Insurance and see what the server does with the various form submission values.

CREATING THE THANKS.ASP PAGE

Suppose our friend Nancy Davolios of the mythical Northwind Trading Company wants a quote on her life insurance, and wants it delivered via e-mail. She might fill in the Humongous Insurance form as shown in Figure 17.15.

Pressing the Submit image button will fire the sendForm() procedure shown earlier. The sendForm() procedure submits the data entered on the Humongous Insurance form. Because its action attribute is set to thanks.asp, ASP will return a custom "Thank You" response page automatically, as shown in Figure 17.16.

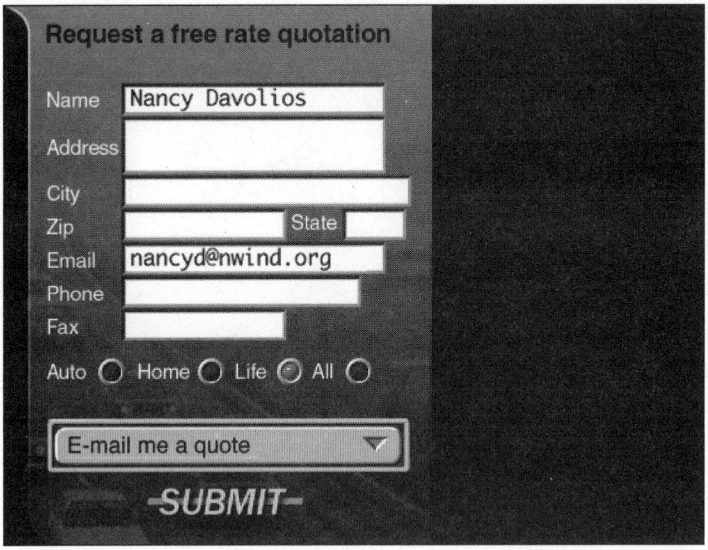

Figure 17.15 *The humform.html page in design mode on the personal computer.*

Figure 17.16 *The thanks.asp return page.*

USING ASP WITH HTML TAGS

ASP script is placed within angle brackets with percent signs in them: <% *script here* %>. If an equal sign is added (<%= *myvariable* %>), variables or the result of scripted procedures can be rendered. The native script for ASP pages is VBScript, but these pages may be written in JavaScript as well. VBScript is used in all the ASP server-side script in this book. Those who prefer JavaScript may include the following line at the top of the page and use JavaScript with ASP.

```
<% Language = JScript %>
```

The key thing to understand about ASP is that it runs on the server side. The form script shown above runs on the machine with the browser on it—i.e., the client side. Since ASP runs on the server, where pages are made, it can be used to create pages on the fly that are customized to any particular client. Values submitted by forms are handled by a built-in ASP object called the *request* object. The request object has a *form* collection that holds values submitted by forms.

Earlier in this chapter, we emphasized the importance of naming form elements, and now you will see why. The value of a form element is referenced using its *name* attribute. For example, Ms. Davolios' name is submitted by the form-input element named txtLastName on the humform.html page, as shown in Figure 17.17.

```
<input type="text" size=20 name="txtLastName">
```

This input will be processed by ASP as follows.

```
Hello <%= Request.Form("txtLastName")%> <br><br>
```

Figure 17.17 *Mixed text, ASP, and HTML gives a custom response.*

Notice that the word "Hello" before the text "Ms. Nancy Davolios" came from ASP script, and the line break was accomplished with standard HTML <br> tags. This mixing of text HTML and server-side script is what makes ASP powerful.

Server-Side Data Validation Using ASP

Users can be counted on to make mistakes when they enter data. For example, they sometimes enter text instead of numbers and numbers instead of text. In addition, they sometimes leave critical fields blank. Data entry errors can be minimized by using Select Option Groups whenever possible. But even this does not ensure correct and relevant data. Data validation is required. Data validation, which can be performed on the client or the server, checks user input for valid data and completeness.

Keyboard events like *onKeypress*, *onKeyup*, and *onKeydown* are not supported on this first release of Microsoft TV. This means a common method of client-side validation is not available to developers of applications for Microsoft TV 1.0. Inline data validation takes advantage of the *onKeypress* event to check input as it is entered. For example, the following example checks input for a field named txtLastName.

```
<input type="text" size=20 name="txtLastName"  onkeypress="valName()" >

function valName()
{
   if((event.keyCode < 65) && (event.keyCode != 32))
   {
      event.returnValue = false;
   }
}
```

As keystrokes are pressed, the *onKeypress* event fires a function named valName. The valName function checks the keyboard event against ASCII values. Improper keystrokes are nullified by setting a return value of false.

```
event.returnValue = false;
```

If keyboard events were supported, this sort of validation could be used on all the fields in a form. Since they are not, other events like *onBlur*, *onChange*, or *onClick* could be used to fire client-side JavaScript data validation procedures. In fact, the *onClick* event used to fire a form submit procedure is a good place to house client-side validation code.

The trouble is there is no such thing as simple JavaScript data validation code. Whole chapters are devoted to this topic in JavaScript books. However, it is beyond the scope of this book to cover client-side data validation using JavaScript.

The thanks.asp page has branching code to handle most of the ways a TV viewer might incorrectly enter data. In most Microsoft Windows programming, if a viewer fails to enter a vital bit of information, the programmer can put up a message or alert box. In a Windows personal computer client-side program, a message like the one shown in Figure 17.18 would come up if a viewer clicked the Submit image button without entering any values in the form.

Figure 17.18 *Windows message box.*

Message boxes like this one cannot be used in either interactive TV or ASP programming. In ASP, the message box would appear on the server, where no one would be available to push the OK button. Programmers should always remember that Microsoft TV cannot put up message boxes in tv mode, which is the mode in which interactive TV programming happens. Similarly, message boxes can not be used with ASP code because there is no one on the server side to push the OK button. The solution to the problem is to use ASP condition blocks with standard HTML to provide cues to viewer.

For example, if the viewer clicks the Submit button on the Humongous Insurance form without giving any information in the form fields, the thanks.asp page is rendered as shown in Figure 17.19.

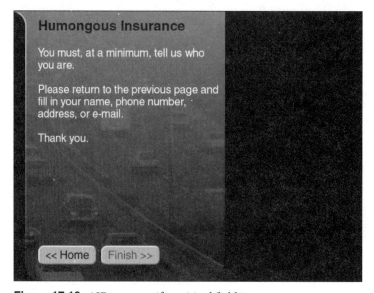

Figure 17.19 *ASP response if a critical field is empty.*

The screen shown in Figure 17.19 was rendered by the following combination of ASP and HTML on thanks.asp:

```
<% Dim strFinish
strFinish = "No"

'If the user did not enter a name,
'give useful directions

    If (Request.Form("txtLastName") = "") Then %>
```

```
You must, at a minimum, tell us who you are. <br><br>
Please return to the previous page and fill in your name,
phone number, address, or e-mail. <br><br>
Thank you.

<% Else %>

<!-- Begin code to handle form input. -->
```

In Figure 17.19, prompting text was rendered in straight HTML and the Finish button was disabled. Because the viewer entered nothing in the txtLastname field on the humform.html page, the item of the same name is empty in the ASP *Request.Form* array. Accordingly, the following condition returns true:

```
If (Request.Form("txtLastName") = "") Then
```

If text had been entered, the Else branch would have been entered and code designed to handle the rest of the form elements would have been used. Next, notice that the Finish button is disabled. It is a good policy to disable buttons when viewers should not use them. We will show how the Finish button is enabled for use when proper data is submitted and then move on to more of the ASP code.

Enabling Controls

The code that enables the Finish button is nearly self-explanatory:

```
<form>
   <input type="button" name="btnHome" value="<< Home"
   onclick="goback()">
   <%If strFinish = "No" Then %>
      <input type="button" name="btnFinish" value="Finish >>"
      disabled="true">
   <%Else%>
      <input type="button" name="btnFinish" value="Finish >>">
   <%End If%>

</form>
```

The variable strFinish is initially set to "No." If it is set to "Yes" in code, as in this example, the Else branch of the If statement is entered and the enabled version of the Finish button (disabled attribute set to "true") is rendered. Notice how elements of the If Else decision block may be broken up with HTML simply by closing the ASP bracket (%>).

In a flushed out example, ASP would render something like "Is this information correct?" regarding the viewer's input. The viewer would indicate whether the information was correct, which would enable the Finish button. By pressing the Finish button, the viewer would send the data to a data store on the server.

ASP Request.Form and Response.Write Methods

The Select Case construct is used to ensure a match between the choices given and information entered. For example, if the viewer chose "E-mail me a quote" from the Select Option Group, the code branch shown below would test whether a value had been entered in the txtEmail input text box. If a value had in fact been entered, the Response.Write method would render proper content on the client "Thank You" page.

> **NOTE** Those wondering whether the specification issued by the ATVEF permitted us to use Select Case instead of If Else should see the section "You Cannot Use Select Case in ECMAScript!" at the end of this chapter.

```
strType = Request.Form("selContactMode")
strFinish = "No"
Select Case strType

   Case "e-mail"
      Response.Write("E-Mail:  ")
      Response.Write("<br>")
      If Request.Form("txtEmail") = "" Then
         strString = "Please return to the form and →
             give us your e-mail address so →
             we may contact you"
         Response.Write(strString)
      Else
         Response.Write(Request.Form("txtEmail"))
         strFinish = "Yes"
      End If
```

If the viewer chose to be contacted via e-mail but forgot to enter an e-mail address, a screen like the one shown in Figure 17.20 is generated.

```
strType = Request.Form("selContactMode")
```

The value of the variable strType holds the value submitted with the form on humform.html from the Select Option Group named selContactMode. In this example, it will be "e-mail", so the case is satisfied and entered.

```
Case "e-mail"
   Response.Write("E-Mail:  ")
   Response.Write("<br>")
```

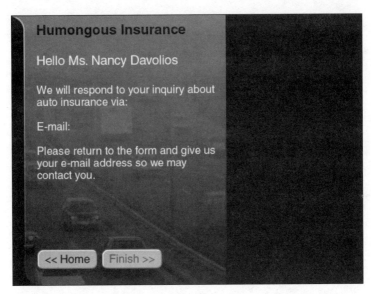

Figure 17.20 *ASP branching can create dynamic input cues.*

The Response.Write method is used to write the literal string of characters in "E-Mail." Next, it is used again to render the HTML tag <br>, thereby forcing a line break. This ability of ASP to render either text or HTML makes its flexibility nearly limitless.

Finally the Request.Form() collection is polled to see if content was sent from the txtEmail input box. If text was entered, the input is rendered on screen and the strFinish variable is set to Yes. The strFinish variable is used to set the disabled attribute of the Finish button. In the next section, we will explain how the Finish button is enabled.

```
If Request.Form("txtEmail") = "" Then
    strString = "Please return to the form and →
        give us your e-mail address so →
        we may contact you"
    Response.Write(strString)
Else
    Response.Write(Request.Form("txtEmail"))
    strFinish = "Yes"
```

Notice that in this example, the Response.Write method can render text within double quotes or any value held by a variable like strString. Setting the strFinish variable to Yes enables the Finish button, because if an e-mail address is entered it is possible to honor the viewer's request, as shown in Figure 17.21. Of course, no validation has been done to see if a valid e-mail address was entered.

When you build more complex validation code remember that it is being run on the server side, so handy VBS functions like *Len, Ltrim, Mid, Rtrim,* and *InStr* may be used to parse strings and develop validation procedures.

VBScript functions are not available for client-side validation code in Microsoft TV 1.0.

Figure 17.21 *Finish button is enabled if it is appropriate.*

In this example the Finish button is vestigial. It only exists as a tool to illustrate the disabling and enabling of controls in ASP code.

Always Use <form> Tags in Microsoft TV 1.0

Internet Explorer–based clients will render a button with or without form tags. Developers can take advantage of this to use form buttons in non-form situations. Since input buttons usually call a procedure for their *onClick* event, the form is optional on Internet Explorer–based personal computer clients.

An important question is: Why is it necessary to wrap the input buttons *btnFinish* and *btnHome* in a set of <form> </form> tags, given that no method or action has been specified? The answer illustrates one of the quirks in the Microsoft TV 1.0 browser. In this browser, form elements are not rendered unless they are enclosed within form tags. For example, the following line of HTML will render a button in Internet Explorer, resulting in an empty space on Microsoft TV.

```
<input type="button" name="btnSomeButton" value="Do That >>"
onclick="myfunction()">
```

The rest of the branches in the Select Case statement handle the various ways that viewers might mistakenly enter data. They all work in the same manner as the case e-mail branch.

You Cannot Use Select Case in ECMAScript!

As previous chapters have mentioned, the specification adopted by the ATVEF committee presents a restriction in regard to the use of Select Case and Switch statements. According to this statement, ECMAScript supports If statements but does not support such constructs as Select Case. It might seem that the demo shown above for Humongous Insurance violates that specification. Nevertheless, the demo is compliant with the ATVEF, because ASP script runs on the server, and ATVEF only applies to the client. Convenient, isn't it?

FOR MORE INFORMATION

In this section, we've described how to lay out form elements to be used for interactive TV. We have also introduced the subject of responding to viewer input. For more information about how to write form data to database back ends or about how to use hidden input elements to pass values between pages, read chapters 18 through 20.

WHAT'S NEXT

In the next chapter you will be introduced to Bob's Pizza. The Bob's Pizza demo illustrates how an interactive TV application can be used as an e-commerce shopping basket where users can choose from a set of products and order online. Order data in Bob's Pizza is actually written to a back-end database using ADO-ASP, Open Database Connectivity (ODBC), and HTML coding.

Chapter 18

Creating Bob's Pizza

In This Chapter

- Where to Find Sample Content for This Chapter

- Introducing Bob's Pizza

- Overview of the Content in This Chapter

- Setting Up the Database DSN

- Setting Up the Desktop Web Server

- Concepts and Techniques

E-commerce—interactive, database-driven commerce over the Internet—has long been the Holy Grail of Web site development. Using the Bob's Pizza demo as a guide, we will start to demystify the process by which database back ends, like Microsoft Access and Microsoft SQL Server, are connected to Web site–based forms viewed on Microsoft TV–based Internet terminal receivers.

Bob's Pizza is the most complex application in this book, so it takes up the most space. In the next few chapters, you will learn how to create Web applications that accumulate customer orders and write them to a back-end database. The pages of the Bob's Pizza demo are optimized for television, but the text in this book should help you understand the underlying concepts. We will focus on the e-commerce aspects of a Web application. The ideas you take away from the Bob's Pizza demo should be useful in building any medium-traffic multiproduct e-commerce Web site, whether it is optimized for TV or not.

Later in this chapter, we will explain how to set up the Bob's Pizza demo on your development system. We will also present an overview of what the various pages in the application do. Chapters 19 and 20 will address client-side scripting, the form elements required to create a pizza order, and the Active Server Pages (ASP) and Microsoft ActiveX Data Objects (ADO) code used to write multiple-line "shopping cart" orders to a database back end.

WHERE TO FIND SAMPLE CONTENT FOR THIS CHAPTER

To get the most out of the material in this chapter, we recommend that you view the sample files on the *Building Interactive Entertainment and E-Commerce Content for Microsoft TV* companion CD and Web site. Many of the pages that make up this application require a Web server. Before you can actually run Bob's Pizza, you will need to place the files in the wwwroot folder of your development workstation Web server and set up a Data Source Name (DSN) for the database that holds order information. Setting up a DSN and running a desktop Web server are topics that have been discussed elsewhere in this book, but we will briefly review them here.

- To view the Bob's Pizza content on a set-top box running Microsoft TV or on the Web, go to *www.microsoft.com/tv/itvsamples*.

- The HTML and database files for Bob's Pizza are located in the "Microsoft TV E-Commerce" section of the companion CD.

- To view the sample content associated with this chapter, including the Bob's Pizza database, see the "Bob's Pizza" topic of the "Microsoft TV E-Commerce" section on the companion CD.

The opening page in Bob's Pizza will resemble Figure 18.1 when displayed in the Microsoft WebTV 2.0 Viewer using an interactive TV link. You will get a reminder of how to create interactive TV links later in this chapter, or you may wish to review Chapter 16, "Creating Interactive TV Links," for a full discussion of that topic.

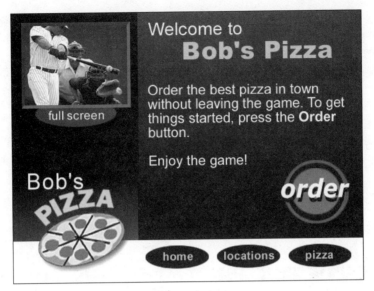

Figure 18.1 *Bob's Pizza in TV mode.*

INTRODUCING BOB'S PIZZA

Bob's Pizza, a fictitious restaurant, is the basis for the interactive TV example described in the next few chapters. The Web pages comprising Bob's Pizza enhance the value of television broadcasts sponsored by companies that can accept product orders from viewers. Bob's Pizza stands to profit from making it fun and easy for viewers to order a pizza. Likewise, the show's viewers are more likely to place an order if they can do so without missing much of the show. Bob's Pizza fulfills both of these requirements.

Bob's Pizza differs from other ads we have presented in that it uses forms that make it possible to order pizzas online while watching a TV show. This characteristic—that is, the combination of forms with interactive TV—gives Bob's Pizza an edge over competitors who use standard "phone me" advertising. It also enables Bob's Pizza to accept and track far more orders than would be possible with a manual "phone it in/write it down" approach.

Forms allow a viewer to exchange information with a database and a Web server. In the case of the Bob's Pizza example, a Web server provides viewers with the necessary information (such as pizza types, sizes, and prices) for placing an order. The viewer then submits the order to the database. From there, Bob's Pizza processes the order and turns it into piping hot pizza ready for delivery.

Getting Web pages connected to a database on a server is not as hard as one might think. That said, the task is no beginner's hayride either. The concepts involved include relational database design, Open Database Connectivity (ODBC), ASP, ADO, server-side scripting in Microsoft Visual Basic Scripting (VBScript) Edition, and client-side JavaScript validation code. To ease your way through this multitude of technologies, we are presenting the essential concepts in bite-size chunks throughout Chapters 18–20. This chapter will focus on the pages used in Bob's Pizza.

> **NOTE** Although the pages are designed specifically for a pizza shop, the concepts and code behind them can easily be applied to other Web-based e-commerce situations. Feel free to copy, change, or otherwise use the code.

OVERVIEW OF THE CONTENT IN THIS CHAPTER

This chapter gives an overview of the pages in the Bob's Pizza demo. It also reviews some of the files that make the Humongous Insurance demo work. These files were constructed using dimensions and techniques covered in earlier chapters, so we will only briefly discuss their structure here. The file descriptions will help you understand how the pages interact to form a complete interactive TV application. Feel free to use or change the files in this demo for your own purposes.

The Bob's Pizza demo consists of the following Web pages:

- **base.html**—a page that defines the nested frameset and the frames for the Bob's Pizza demo.

- **tvlogo.html**—a page that defines the Web page in the left frame. This page contains the TV control, the logo, and advertising. The tvlogo.html page is always visible in the left frame.

- **home.html**—an introductory page that appears in the right frame on Open.

- **locate.html**—a page that fills the upper right content frame if the Order or Locations button is clicked with the home page in view.

- **specials.html**—a page that fills the content area after a location has been selected or the Pizza button in the lower right frame has been clicked. This is the crucial form page of Bob's Pizza. Users assemble their pizza orders by selecting the type, size, and quantity on specials.html. This page also presents the pizza "specials" for each day.

- **order.asp**—a page that fills the content area after the Order button is clicked on the specials.html page, indicating that a viewer decision has been made about the type, size, and quantity of pizza. This page allows users to assemble multiple pizza types and edit their order before committing it to the database.

- **customer.asp**—a page that fills the content area when the Next button on order.asp is clicked, indicating that the customer has assembled an order and is ready to complete the transaction. A form is displayed to obtain the customer information needed to complete the order, deliver the pizza, and receive payment.

- **thanks.asp**—a page dynamically created by ASP technology in response to the submit event on customer.asp. Several technologies, including SQL string language, ADO, ASP, and VBScript, are used to combine and display the user's choices and give information about the order. Batch updating is employed to write order data to several relational tables in the Access back-end database.

- **adovbs.inc**—a special file used to provide familiar VBScript constants such as *adOpenKeyset* to a VBScript environment, such as an ASP-ADO Web page.

- **cssPC.css and cssTV.css**—cascading style sheets (CSS) used to render proper fonts to the platform, as described in Chapter 23, "CSS Level 1 Support for Microsoft TV."

- **viewset.html**—a test page that acts as a proxy for the broadcast link that will be sent along with a video signal on line 21 of the vertical blanking interval (VBI). The VBI and other issues surrounding the transmission of interactive content to the TV client were covered in Chapter 15, "Fundamentals of Delivering Interactive TV Content," and Chapter 16, "Creating Interactive TV Links."

SETTING UP THE DATABASE DSN

The order form area in Bob's Pizza reads and writes to a Microsoft Access database. It could just as easily communicate to a SQL Server or another database management system. Access was chosen for this demo because of its ubiquitous deployment, ease of use, and portability.

The first step in facilitating data exchange between a server and a Web page form is to set up an ODBC driver for the Bob's Pizza database, called a Data Source

Name (DSN). Just as your printer needs a driver to bridge the language barrier between its internal functionality and your computer, database applications use drivers to communicate between Web page code and their underlying functionality. All the major database management systems (DBMS), such as Access, SQL Server, Oracle, Sybase, and others, have ODBC drivers built for their database systems and designed for purposes such as this. You do not even have to go looking for these ODBC drivers, because most of them are included with Microsoft Windows.

> **NOTE** ODBC and DSN are not just Web e-commerce solution technologies; they are extensively used in enterprise-level office automation and client-server data-sharing applications. Just as with printer or video drivers, it is a good idea to periodically check the Web site of your DBMS vendor for updated ODBC drivers.

To set up the ODBC driver for Bob's Pizza DSN:

1. On the personal computer running the server that you plan to use as a desktop Web server (Windows NT Server, Windows 2000 Server, or Windows 98), create a new folder called BobsPizzaData.

2. Insert the companion CD for this book into the CD drive and copy the Bobs.mdb file from the Bobs_Database folder in the "Microsoft TV E-Commerce" section of the companion CD to the BobsPizzaData folder on the personal computer.

3. Click Start, point to Settings, and then click Control Panel. On machines running Windows 2000, click Administrative Tools to expose the Data Sources OBDC icon.

4. To open the ODBC Data Source Administrator dialog, double-click the ODBC icon shown in Figure 18.2.

ODBC

Figure 18.2 *ODBC icon in Windows Control Panel.*

5. To open the Create New Data Source dialog as seen in Figure 18.3, click on the System DSN tab and then click the Add button in the ODBC Data Source Administrator dialog.

6. To open the ODBC Microsoft Access Setup dialog, select Microsoft Access Driver (*.mdb) from the list and then click Finish in the Create New Data Source dialog.

NOTE Windows 2000 dialog boxes may vary slightly from the figures shown.

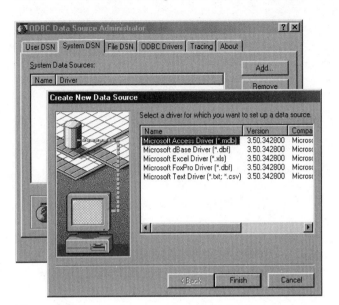

Figure 18.3 *Create New Data Source dialog box.*

7. To complete the creation of a DSN for Bob's Pizza, type Bobs in the Data Source Name in the ODBC Microsoft Access Setup dialog. In the Description box, type **Sample database for Bob's Pizza**. Click Select and locate the Bobs.mdb file in the BobsPizzaData folder that you created earlier. The ODBC Microsoft Access Setup dialog box should resemble Figure 18.4.

Figure 18.4 *ODBC Microsoft Access 97 Setup dialog box.*

That is all there is to setting up a DSN. You may wish to use the same steps to set up a DSN to one of your own database back ends. The DSN will enable us to identify which database our ADO code will read from and write to.

The next section of this chapter covers setting up files to run on a desktop Web server. If you already did this for the Humongous Insurance example in Chapter 17, "Creating Forms for Microsoft TV Content," you may want to only skim the next section.

SETTING UP THE DESKTOP WEB SERVER

Most of the demo applications in this book may be viewed right off the CD-ROM or your hard drive in a standard Web browser like Microsoft Internet Explorer, or in the WebTV 2.0 Viewer. The pages that make up Bob's Pizza require a bit more computer power than standard HTML pages. By power, we do not mean more memory or megahertz; rather, these pages require Web server capability. Bob's Pizza requires a Web server in order to run ASP along with server-side script; it also needs a Web server to run ADO code, which is necessary to write form data to the Access back-end database.

If you have not already done so, you will need to install either Personal Web Server (PWS) or Internet Information Services (IIS) on one of the machines in your interactive TV development workstation. Chapters 3 and 17 provide instructions for setting up and using a Web server on your desktop.

> **NOTE**   PWS comes free with Windows 98, but it is not part of the default installation of Windows. IIS is installed with Windows NT Option Pack 4 (where it is named Internet Information Server) and with Windows 2000. Both PWS and IIS come with the ADO-ASP functionality required for Web database connectivity. To download Service Pack 4, go to *http://www.microsoft.com/support/win\nt/default.htm*.

If you are unsure about your Web server status, open Windows Explorer and look for a folder named Inetpub. If you have Inetpub, you have a Web server.

To view Bob's Pizza in the WebTV Viewer:

1. On the personal computer running Web server software, open Windows Explorer and go to the \Inetpub\wwwroot folder.

2. On the File menu, point to New and then click Folder.

3. Type **BobsPizza** for the name of the folder.

4. Place the CD that accompanies this book into the CD-ROM drive. Copy all the files and images from the Resources folder on the CD to the newly created BobsPizza folder, as shown in Figure 18.5.

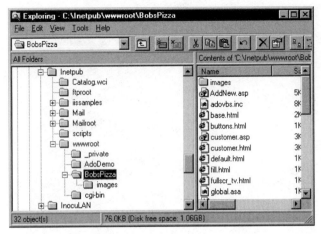

Figure 18.5 *The wwwroot folder with BobsPizza subfolder.*

5. If you have not already done so, install the WebTV 2.0 Viewer using directions given in Chapter 3, "What You Need to Create and Deliver Microsoft TV Content." On a client, open the WebTV 2.0 Viewer.

6. In the URL box of the WebTV 2.0 Viewer, type ***http://ServerName/ BobsPizza/base.html***

7. *ServerName* is the name of the personal computer in your development workstation setup that runs PWS or IIS Web server software.

   NOTE If you are unsure of the Web server personal computer's name, right-click on the Network Neighborhood icon on the personal computer's desktop. Click Properties. On the Identification tab, look in the Computer Name box for the name of the personal computer. If you are using a lone Windows 98 machine with PWS, you are both the client and the server. To act as both client and server, you must loop back the URL value by supplying the IP address in place of machine name. You may determine your IP address by typing **IPCONFIG** in an MS-DOS window at the C:\Windows> prompt. See Chapter 3 or Chapter 17 for detailed instructions on using a single machine with PWS and WebTV Viewer.

Figure 18.6 shows the first page of Bob's Pizza as seen through the WebTV Viewer, using the method described above. The WebTV Viewer does a fair job of rendering content as it will be seen on television, but it is not ideal. If you have the setup recommended in the section of Chapter 3 titled "Recommended Setup for Creating Microsoft TV Content," you should use it. We recommend copying Bob's files up to a Web server that is outside your company firewall, and viewing them on a television through a WebTV Plus receiver. If servers outside the firewall are unavailable to you, you can contract with an Internet service provider (ISP).

Figure 18.6 *Bob's Pizza in WebTV 2.0 Viewer Web mode.*

The representation of the Bob's Pizza Web page shown in Figure 18.6 is easy to accomplish, but it does not show the content in TV mode. To show Bob's Pizza in TV mode, you can create an interactive TV link in the WebTV Viewer. If you are not sure how to create an interactive TV link in the WebTV Viewer, see Chapter 16. Figure 18.7 shows the interactive TV link created for the Bob's Pizza demo as it would appear on a machine acting as both client and server.

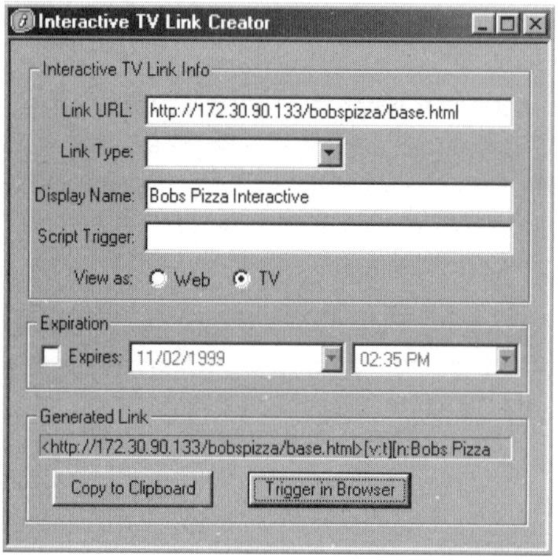

Figure 18.7 *Interactive TV link on a machine acting as both client and server.*

Notice that an IP address is used in place of the more usual computer name as the URL portion of the interactive TV link syntax. This causes the WebTV Viewer to "loop back" and render ASP pages properly. The use of plain computer name syntax on a machine acting as its own Web server will cause the computer to bypass PWS and render ASP pages incorrectly in the WebTV Viewer. Naturally, your IP address will differ from the one shown in Figure 18.7. To determine your IP address, refer to notes given earlier in this chapter or to the more detailed information in Chapters 3 and 17. If you are using a single machine as your Web server and interactive TV client, don't expect blazing performance out of ASP pages. There will be a short lag between the time you click on something and the expected result.

If your development setup contains at least two computers, use one of them as the client and the other as the Web server. With a two-computer development workstation, you may use your computer name in your URL syntax, as shown below:

`http://mywebserver/BobsPizza/base.html`

It is instructive to open Bob's Pizza in the WebTV Viewer or a true WebTV set-top box and cycle through the various pages as they are described here. This process will prove helpful later in this chapter when we explore how Bob's Pizza functions. Figure 18.8 shows Bob's Pizza after the Order button on its opening page has been clicked.

Figure 18.8 *Bob's Pizza locate.html page.*

CONCEPTS AND TECHNIQUES

Basic HTML concepts covered earlier in the book will not be repeated here, except for hints about modifying the example pages for your own use. We strongly recommend that you open the HTML pages that comprise this example (using the text editor of your choice) and look at the code that is discussed in the following sections. The example pages are available in the Templates section of the companion CD.

Setting Page Dimensions in a Nested Frameset

One of the recurring messages in this book is: "Use the design space defined by the limitations of the TV monitor." The design space we have to work with is 560 pixels wide and 420 pixels high. Pages that occupy the design space described by a frameset are limited to 528 pixels in width. The logic behind these values is described in Chapter 5, "Guidelines for Designing Microsoft TV Content."

What's Novel About Bob's Frameset?

Most of the applications in this book have used simple framesets consisting of two columns with a total width of 560 pixels. The frameset that makes Bob's Pizza, however, is nested; that is, the right column of the outer frameset contains a frameset of its own. This inner or nested frameset allows the right part of Bob's Pizza to contain both fixed and changing content.

In Bob's Pizza, the base.html page defines the frameset used in this design space. Open this page in a text editor to see for yourself how this design space works, and to gain insight into the skeletal structure underlying the whole of Bob's Pizza.

```
<frameset border=0 frameborder="0" framespacing="0"
noresize rows="420,*" cols="182,378,*">

<frame name="fralogo" src="tvlogo.html" marginheight=0  noresize
frameborder=0 scrolling="no">
</frame>

<frameset border=0 frameborder=0 noresize framespacing=0 rows="361,49,*">

<frame name="fraorder"  height="361" src="specials.html" noresize
frameborder=0 scrolling="no">
</frame>

<frame name="frabuttons"  height="49" src="buttons.html" noresize
frameborder=0 scrolling="no">
</frame>

</frameset>

</frameset>
```

Notice that the outermost frameset is composed of one big row, which is 420 pixels tall and is split by two columns that total 560 pixels (182 + 378 = 560). The 182-pixel column on the left is the *fralogo* frame and holds the video control. The 378-pixel column on the right is nested into two rows: the bottom row is the *frabuttons* frame, which is 51 pixels high and contains a set of navigational buttons; while the top row is the *fraorder* frame, which is 369 pixels high and houses the interactive forms used by Bob's Pizza (369 + 51 = 420). When taken together, these dimensions create a frameset that enables the pages of Bob's Pizza to exist within the limitations imposed by a TV monitor. Remember that content pages must not exceed 528 pixels in width or the pages will lack proper margins.

NOTE It is important to give frames a name so that they can be acted on in script. This concept is addressed in more detail in Chapter 20, " Creating ASP-ADO Code to Interface with the Database."

The Contents of Bob's Frameset

This section will describe the various HTML and ASP pages that populate the parent frameset of Bob's Pizza.

fralogo—The Video Object Frame

Control of the *fralogo* frame is accomplished through the tvlogo.html page. This page also contains the coding for the TV object and the Full Screen button. The tvlogo.html page is shown in Figure 18.9.

Figure 18.9 *The tvlogo.html page in Web mode.*

Files used in the *fralogo* frame:

■ tvlogo.html contains the logo image and video control.

The width of the *fralogo* frame (182 pixels) is determined by the widest element that must be accommodated in the frame, which in this case is the Bob's Pizza logo. If you adapt this example for your own use, the easiest path will be to create your logo within this 182-pixel envelope. If you deviate much from this constraint, the other elements of the page will fall out of alignment.

Another issue to be aware of when modifying this example for your own use is that some of the positioning of elements on the tvlogo.html page is accomplished using inline attributes of the CSS Style object within <div> tags. For example, the following line of dynamic HTML (DHTML) code positions a table containing a video control and the Full Screen button on the page:

```
<div id="video" style="position:absolute; left:5; top:15">
```

This sort of positioning code works both in WebTV Plus and Microsoft TV–based Internet receivers, but it may not work in a receiver built to a minimal interpretation of the Advanced Television Enhancement Forum (ATVEF) Specification for Interactive Television 1.1. If you anticipate that your viewer market contains receivers that comply with the specification adopted by the ATVEF but do not support CSS positioning, you must use the <table> tag to do any fine positioning of elements.

NOTE Currently, the TV receivers that we are aware of include at least some CSS positioning functionality. For complete information on what a minimal interpretation of the specification adopted by the ATVEF means in terms of CSS properties, see the "Microsoft TV Programmer's Guide" contained on the companion CD.

frabuttons—The Navigational Buttons Frame

Files used in the *frabuttons* frame:

■ buttons.html

Just as the logo for Bob's Pizza determined column width, the height of the button images determines the row height for the *frabuttons* frame. If you need to change these button images for your own use, your best bet is to maintain their current dimensions of 160 pixels wide and 40 pixels high. Alternately, you could use standard HTML buttons instead, if you prefer not to work with graphical images. The following code sample demonstrates how you can implement an HTML button that, when clicked, calls a function:

```
<input type="button" name="mybutton" onclick="somefunction()">
```

The *frabuttons* frame contains a very simple page named buttons.html, as shown in Figure 18.10.

Figure 18.10 *Navigation buttons in the* frabuttons *frame.*

Predictably, buttons.html contains the buttons used to load new content into its larger counterpart, the *fraorder* frame. The Pizza button loads the specials.html page, which is an order form for the various pizza specials. The Home button loads the home.html page, and the Locations button loads the locate.html page into the *fraorder* frame.

fraorder—The Order Form Frame

The *fraorder* frame takes up the entire upper right quadrant of Bob's Pizza. It is the star of the show, because *fraorder* is where all the action takes place. Several different pages occupy *fraorder* during the course of ordering a pizza.

Files used in the *fraorder* frame:

- **home.html**—the opening occupant of this frame.

- **locate.html**—a page that appears if the Order button on home.html or the Locations button on buttons.html is clicked.

- **specials.html**—a page that fills the *fraorder* frame if either the Locations button or the Pizza button is clicked.

- **order.asp**—a page that loads as a response to the Submit event on specials.html.

- **customer.html**—a page that appears after the user has assembled an entire order.

- **thanks.asp**—the final occupant of *fraorder*. It is triggered by the submit event of a form on customer.asp.

> **NOTE** The Bob's Pizza home and location pages are cosmetic and explain themselves. You may wish to use opening pages like these if they in some way empower your application. For example, users of Bob's Pizza may improve their order response time by choosing a nearby location.

The specials.html Page

The specials.html page contains the form elements that enable viewers to order the pizza "specials" in various size and quantity configurations, as shown in Figure 18.11.

Figure 18.11 *The specials.html form page in WebTV Viewer in TV mode.*

Try playing with some of the form elements in specials.html. Notice that the pizza description changes when you select a different type of pizza, and price totals update when you select a new pizza size or quantity. For example, Figure 18.12 shows the result of choosing two 16-inch "veggie" pizzas.

Figure 18.12 *Client-side JavaScript calculates order values.*

Notice that each change in the pizza-type Select Option Group changes the description in the Pizza Description text area and the price in the Price text box. Similarly, changing pizza size using the Size Radio Button Group, or changing the number of units using the Quantity Select Option Group, changes the price. Both the text area and the Price text box on the Specials form are inaccessible to user focus by design.

A Select Option Group, rather than a text box, was chosen for quantity input because Bob would not want to receive a gag order for 999 pizzas. In general, Select

Option Groups are preferable to text boxes for interactive TV forms. This is because option groups may be selected using the WebTV infrared remote control, unlike text boxes, which require typing. Even though most set-top box owners will have an infrared keyboard and others may enter text using a virtual keyboard, nearly all owners will find that typing is a pain in Microsoft TV. Sometimes text boxes are the only solution, but developers should do all they can to prevent viewers from needing to resort to their keyboards frequently.

NOTE   Instructions for detecting user information can be found on the subscription database that goes with every set-top box. Look at *http://www.microsoft.com/ tv* for examples of code that will use cached information to fill out forms for viewers.

The code that makes updating possible is covered in the next two chapters. For now, we recommend that you experiment with the various combinations of pizza type, size, and quantity to get a feel for the way things operate.

When you have played with specials.html as much as you like, click the Order button to open the first truly database-driven page, order.asp.

The order.asp Page

Upon clicking the Order button on the specials.html page, the order.asp page pictured in Figure 18.13 loads into the *fraorder* frame. This ASP page is dynamically built, using information entered by the viewer on the specials.html page. It acts as a "shopping cart" for Bob's Pizza.

To begin adding items to the shopping cart, choose some combination of pizza size, type, and quantity from the Specials form. Figure 18.13 shows the result of the order keyed into the page shown in Figure 18.12.

Figure 18.13 *The order.asp page is a dynamically-built ASP "shopping cart."*

From the order.asp page, the viewer can add more pizzas by clicking the Add button, which returns to the specials.html page where additional pizzas can be selected. Alternatively, the viewer can remove the pizza currently selected in the Order Box by clicking Remove Item. Thus, viewers can add and remove pizzas until the order size and price fits their snacking requirements. For example, if the viewer returned to the Specials form and added a 12-inch kahuna pizza to her order, the order.asp Order Box would resemble Figure 18.14.

Figure 18.14 *The order.asp page with three pizzas ready for order.*

Notice how both the 16-inch veggies and the 12-inch kahuna are now in the Order Box. The Bob's Pizza application was able to maintain state for the variables that make up this user's order, even though the pages that filled them closed and opened again. The remarkable ability of ASP to maintain state across pages even works if the viewer changes television channels and returns!

The next chapter will explain what it means to dynamically build a page. In a nutshell, ADO code is used to open a record set on a table named "shopcart" in the back-end Access database. The Shopcart table contains information about each order item, such as the type of pizza and the quantity ordered. A field named SessionID that uniquely identifies each particular user makes it possible to combine the selections by the user. ASP ships with a special object named the Session object. It has a method, Session.SessionID, that generates a number unique to each user and lasts for 20 minutes. This unique session number acts like a server-side cookie, enabling Web developers to track a user across many pages without affecting the user's hard drive.

You can see this tracking in action by opening Bob's Pizza on two separate machines. This will create two session i.d. numbers, and you can fill two "shopping carts" with various types and quantities of pizza, as if there were multiple users logged onto the Web site. We suggest that you open the Shopcart table in the Access database that you set up as the DSN for this application. This will give you a sense of how orders are accumulated. See Figure 18.15 for a look at how the Shopcart table with this order is stored temporarily until the viewer finalizes it.

SessionID	PizzaType	PizzaSize	PizzaQuan	PizzaTotal	ProductID
72719533	veggie	16	2	$34.78	9
72719533	kahuna	12	1	$13.39	10
0			0		0

shopcart : Table

Figure 18.15 *Temporary home for order data is the Shopcart table.*

Before moving on, notice that order quantities and total dollars are shown in a table that updates each time the order.asp page is called. To try this out, select one of the items in the Order Box, and then click Remove. The page refreshes itself with the totals updated.

Finally, check out the validation code built into order.asp that keeps users from doing things that would cause errors. For example, an error would pop up if there were no items in the Order Box and the user either tried to move to the next page or kept clicking Remove. To see how this is prevented, remove all the items from the Order Box until you see a screen like the one shown in Figure 18.16.

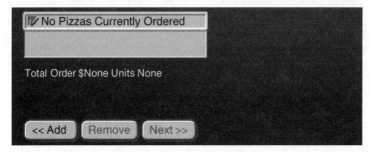

Figure 18.16 *Users cannot cause an error because the Remove and Next buttons are disabled.*

Notice that the Order Box displays a "No Pizzas Currently Ordered" message and that the Remove and Next buttons are disabled. One of the lessons developers learn early is that if there is a way for users to break your code, they will. Validation code, like the code presented in the next two chapters, is essential to the creation of a "bulletproof" interactive TV e-commerce application.

Now that you have seen how orders are assembled, it is time to gather the information necessary for one of Bob's employees to deliver the pizza and get paid.

The customer.asp Page

If there are any pizzas in the Order Box on the order.asp page, the Next button is enabled. Clicking the Next button, as one might expect, takes the user to the next page. By using this paged wizardlike format, Web applications can gather enough information to complete an order without trying to cram too much information on a single page. This is very important in the scroll-free world of Microsoft TV–based Internet receivers and other set-top boxes.

When the viewer clicks the Next button on the order.asp page, the customer.asp page appears. This page gathers information about who the customer is and where to deliver the pizza. As Figure 18.17 shows, the customer form is simply an HTML form relating to the pizza order.

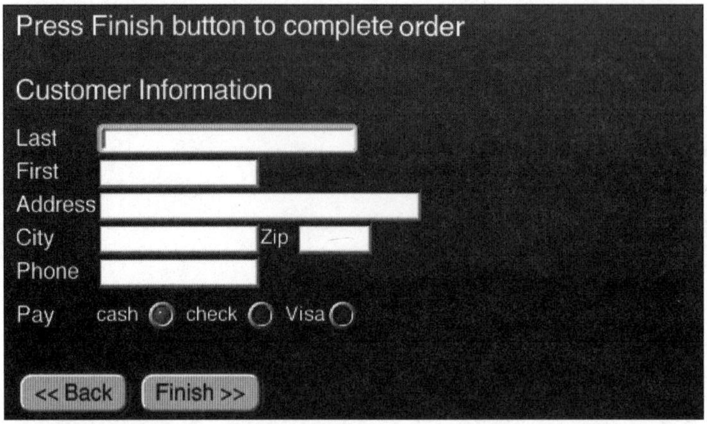

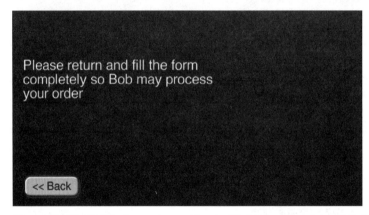

Figure 18.17 *The customer.asp page shown in WebTV for Windows.*

If data entered is insufficient to complete an order, the TV viewer sees a screen like the one shown in Figure 18.18.

Figure 18.18 *Validation code enables the Finish button when all fields are full.*

Clicking the <<Back button returns a user to the form on customer.html.

The thanks.asp Page

The final page in the Bob's Pizza application is thanks.asp. (See Figure 18.19.) This page is the real ASP-ADO workhorse of the application. It compiles all the order and personal information from the user, then opens record sets on the Customer Order and Shopcart tables. It writes records in the tables that make up the back-end database. It then closes the customer.asp form window and generates a personalized "Thank You" page that appears in the *fraorder* frame.

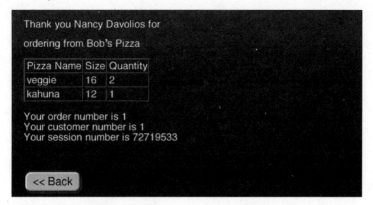

Figure 18.19 *The thanks.asp page shown in the* fraorder *frame.*

The remarkable characteristics of thanks.asp will be revealed in Chapter 20.

WHAT'S NEXT

This chapter offered an overview of how an e-commerce site such as Bob's Pizza works. The next chapter will focus on creating the first order entry page for Bob's Pizza—specials.html. While not an ASP page, specials.html introduces many valuable techniques, including setting up a form, choosing a method of submitting data to the database, using dynamic form controls, and incorporating JavaScript client-side coding.

Building Bob's Order Entry Page

In This Chapter

■ Where to Find Sample Content for This Chapter

■ How specials.html Fits into the Bob's Pizza Demo

■ Using the WebTV Viewer to View Content

■ Code and Controls on specials.html

The previous chapter covered how the Bob's Pizza demo was constructed with a frameset, a series of Web pages, and a back-end database. This chapter focuses on a particular Web page of Bob's Pizza called specials.html. The specials.html page is the order entry page that enables TV viewers to select a pizza and to specify the size and quantity of pizzas they wish to order. We will explore how specials.html was created and demonstrate how the form, form elements, and JavaScript work together to establish the front end of an e-commerce application. After you have read this chapter and mastered the construction of the specials.html page, you will be ready to learn how VBScript is used with Active Server Pages (ASP) and Microsoft ActiveX Data Objects (ADO) to write a pizza order to a Microsoft Access database on a server.

WHERE TO FIND SAMPLE CONTENT FOR THIS CHAPTER

It is strongly recommended that, as you read this chapter, you open the Web pages in a text editor, such as Microsoft Windows Notepad, and follow along as concepts are explained. It is also instructive to view the content covered in this chapter on the Microsoft WebTV Viewer.

- To view the Bob's Pizza content on a set-top box running Microsoft TV or to view the content on the Web using a personal computer, go to *http://www.microsoft.com/tv/itvsamples.*

- The HTML and ASP files that make Bob's Pizza work are located in the Bobs_Demo_Code folder under "Bob's Pizza" in the "Microsoft TV E-Commerce" section of the companion CD.

- Files that explain how to set up and use the Bob's Pizza demo are located in the Bobs_Readme folder. The database that goes with this application is located in the Bobs_Database folder.

HOW SPECIALS.HTML FITS INTO THE BOB'S PIZZA DEMO

As discussed in the previous chapter, Bob's Pizza is a frames-based application that consists of two frames. The left frame, called *fralogo,* holds the tvlogo.html page that contains the TV object and the Bob's Pizza logo. The right frame, called *fraorder,* holds the specials.html page when the Bob's frameset first loads, as shown in Figure 19.1.

Figure 19.1 *Bob's Pizza.*

USING THE WEBTV VIEWER TO VIEW CONTENT

If you have read Chapter 18, "Creating Bob's Pizza," you may already be set up with Internet Information Services (IIS) or Personal Web Server (PWS). If you want to continue to use your Web server setup to review this chapter's content, feel free to do so. However, a Web server is not required for this chapter because the specials.html page covered here is built with standard HTML, so it can be opened directly from your hard drive or from the CD that accompanies this book.

Copy the files of Bob's Pizza:

If you have not already done so, create a folder on your hard drive, and name it BobsPizza. Copy the files and images from the Bobs_Demo_Code folder under "Bob's Pizza" in the "Microsoft TV E-Commerce" section of the companion CD to the BobsPizza folder on your hard drive.

Now, open the specials.html page in the WebTV Viewer.

Open Bob's Pizza in the WebTV Viewer:

1. Open the WebTV Viewer.

2. From the File menu, click Open File.

3. Browse to the BobsPizza folder and click on specials.html. The "specials" page loads in the WebTV Viewer, as shown in Figure 19.2.

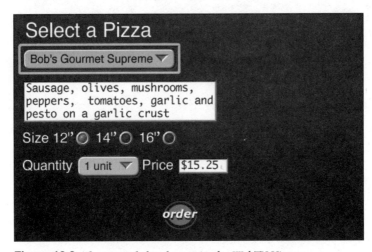

Figure 19.2 *The specials.html page in the WebTV Viewer.*

As you may have noticed, when you load the specials.html page into the WebTV Viewer, form elements such as radio buttons and Select Option Groups render differently in the television environment than they do on a personal computer. That is why the WebTV Viewer is such a handy tool. It reveals what you can expect to see on TV while you are working on a personal computer. Now that you have seen how the specials.html page will look when shown on TV, let's look at the HTML code and JavaScript used to construct the page.

CODE AND CONTROLS ON SPECIALS.HTML

The specials.html page provides a series of interrelated form elements. Before delving into a technical discussion about forms, form elements, and JavaScript, take a few minutes to play with the form controls to see how the form works. Notice that as you change the value of the controls, the value in the Price box automatically updates to reflect the selections. Also notice that when the type of pizza is changed, the description area below the selection changes as well, as shown in Figure 19.3.

Figure 19.3 *Form elements on the specials.html page.*

Form Elements on specials.html

To create the form for the specials.html page, a variety of form elements are used, including:

■   **<form></form>**—Form elements must be placed within form tags to be visible in Microsoft TV.

■   **<select></select>**—Select Option Groups are used to specify the type and quantity of pizza ordered. Select Option Groups are the best way to give users keyboard-free choices in Microsoft TV applications.

■   **<option></option>**—Option tags create the choices within the <select></select> tag sets.

- **\<textarea>\</textarea>**—Textarea tags are used to create areas where multiple lines of text can be dynamically updated. The textarea tag is used twice. The first instance holds descriptive information about pizzas for sale. The second textarea tag set holds calculated pizza price values.

- **\<input type=radio>**—Radio buttons are used for selecting the pizza size.

- **\<input type=hidden>**—The hidden input type is used to pass user choice information to the next page of the application. Hidden fields, as the name implies, are not visible to the TV viewer.

HTML Form Code on specials.html

Now that we have reviewed the types of form elements used for the specials.html page, let's look at the HTML code used to construct the page. You can view the HTML code in your HTML editor, or you can view the code in Microsoft Internet Explorer.

To view the source code of specials.html in Internet Explorer:

1. In the BobsPizza folder on your hard drive, double-click specials.html. The page opens in Internet Explorer or in your default Web browser.

2. In the View menu of Internet Explorer, click Source. You should now see the source code as shown in the following example:

```
<form name="frmPizza" method="post" action="order.asp">
<!--option group of pizza specials-->

<select name="selPizza" style="position:relative; left:5"
onChange="calcprice()">
<option value="gourmet" tabindex=0> Bob's Gourmet Supreme
<option value ="veggie"> Garden Veggie Delight
<option value ="meat"> Meat Eater
<option value ="kahuna"> Big Kahuna Hawaiian
</select>

<br><br>

<textarea name="txadesc" readonly rows=4 cols=30>
Sausage, olives, mushrooms, peppers, tomatoes, garlic
and pesto on a garlic crust
</textarea>

<br><br>
```

(continued)

```
Size
<input type="radio" name="radSize" checked value="12"
onclick="calcprice()"> 12"
<input type ="radio" name="radSize" value="14"
onclick="calcprice()"> 14"
<input type ="radio" name="radSize" value="16"
onclick="calcprice()"> 16"

<br><br>

Quantity
<select name="selQuan" onChange="calcprice()">
<option value=1> 1 unit
<option value=2> 2 units
<option value=3> 3 units
<option value=4> 4 units
<option value=5> 5 units
<option value=6> 6 units
<option value=7> 7 units
<option value=8> 8 units
<option value=9> 9 units
</select>

Price
<input type="textarea" readonly name="txaprice" size=7
value="$15.25">
<!--Set to product number for default pizza = gourmet 12"-->
<input type="hidden" name="hidprodid" value=4>

</form>
```

Attributes of HTML Form Tags

The opening <form> tag of the specials.html page has three very important attributes:

■   **name**—identifies this form in script.

■   **method**—determines how form data is submitted. The two possibilities are *get* and *post*.

■   **action**—determines the page that form data is submitted to.

The following code snippet shows how the name, method, and action attributes are used for the <form> tag of Bob's Pizza:

```
<form name="frmpizza" method="post" action="order.asp">
```

The Name Attribute

The name attribute is used in scripting code to identify the form and its elements. For example, the following script, called by the *onClick* event of the Order button on the specials.html page, submits the form named frmPizza:

```
function myorder()
{
   var thisform = document.forms['frmPizza']
   thisform.submit()
}
```

You can refer to forms by their index number in the *document.forms* array, or you can refer to them by name. Named form elements become properties of the document object. All three of the following lines return a reference to the first form on a page if the form is named frmPizza:

- ```
  var thisform = document.forms['frmPizza']
  ```

- ```
  var thisform = document.forms[0]
  ```

- ```
  var thisform = document.frmPizza
  ```

Implementation of the Document Object Model (DOM) is not complete in Microsoft TV. This will likely be true for other new TV receiver devices built to the guidelines adopted by the Advanced Television Enhancement Forum (ATVEF). If you run into a dead end in your scripting, it is useful to know that in many instances, multiple solutions are available.

The Method Attribute

The method attribute determines how form information is submitted to the server. The two methods are get and post. The get method is favored by Common Gateway Interface (CGI) programmers. Unfortunately, this method has significant security and functionality disadvantages, because user data is transmitted along with the URL and is visible to anyone who might be packet sniffing TCP/IP packets as they flow across the Internet. In the ASP-ADO e-commerce programming used in Bob's Pizza, the post method is employed. The post method transmits data as part of the IP header, so the data is not visible and is therefore more secure.

> NOTE For a more complete discussion of the merits and drawbacks of the get and post methods, see Chapter 17, "Creating Forms for Microsoft TV Content."
  ```

The Action Attribute

The action attribute of a form determines the page that the server should send back to the client. For Bob's Pizza, the *submit* event sends form information from specials.html to the server, where the form data is processed into a new page named order.asp. The order.asp page then replaces specials.html in the *fraorder* frame of the frameset.

> NOTE   For complete information on the base frameset page or other skeletal aspects of the "Bob's Pizza Interactive" Web site, see Chapter 18, "Creating Bob's Pizza."

The order.asp page provides feedback on the number and type of pizzas ordered and the combined price for each order. It also gives the viewer a chance to adjust the order by removing pizzas from the shopping basket or by returning to specials.html to add more pizzas. We will cover order.asp, as shown in Figure 19.4, in more detail in Chapter 20, "Creating ASP-ADO Code to Interface with the Database."

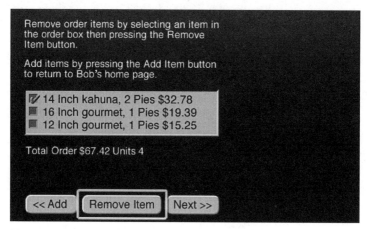

Figure 19.4 The *order.asp page from "Bob's Pizza Interactive."*

For now, let's move forward with an explanation of the form on the specials.html page.

Textarea vs. Plain Text Input Boxes

Textarea controls are used in two places on the specials.html page. The first place is the textarea control named txadesc, which provides text that describes pizza. The next textarea control is txaPrice, which indicates the price of a pizza order. Two characteristics of textarea controls that differ from plain HTML <input type="text"> text controls are important in Bob's Pizza:

- Unlike <input type="text"> text controls, textarea controls may display more than one row of text.

- Unlike a text-input box, the textarea box can be set to *readonly* so that TV viewers cannot change its value. This means that the text in a textarea may be shown like an active control, but TV viewers cannot change it. For input text boxes, you can prevent user interference by disabling the control, but the text in a disabled text box appears grayed out.

Textarea vs. InnerHTML

The other big reason for using textarea controls pertains to limitations of the Microsoft TV browser and restrictions in the specification adopted by the ATVEF. The innerHTML property is often used by developers in situations where text must be changed on the client side using the DOM and JavaScript. InnerHTML is not supported in Microsoft TV 1.0, but it is slated to function in the next release of Microsoft TV.

Textarea vs. DIV Tags

Another possible way of changing text is to use <div> tags containing various messages and then hide and show the DIVs as needed. This method makes use of the *document.all* array and the style property. Although these techniques are supported in Microsoft TV 1.0, they are not part of European Computer Manufacturers Association 262 Language specification (ECMAScript) and hence are not within the guidelines of ATVEF coding. Therefore, if a TV receiver manufacturer implements the specification adopted by the ATVEF in a minimal way, hiding and showing DIVs will not work.

The bottom line is that scripting to textarea controls is currently the safest way to programmatically change text on a page designed for Microsoft TV.

Option Groups and Radio Buttons

Option groups and radio buttons are a good choice for Microsoft TV for several reasons:

- Most likely, TV viewers will have a wireless keyboard, but it is best not to require them to use the keyboard too often. Option groups and radio button choices enable a user to input information with a few simple clicks of the remote control.

- In addition to the usability advantages, option groups make data validation code a snap because users cannot enter invalid information, such as a number where a letter is required.

- Option groups expand without causing the page to scroll, which is important because Microsoft TV does not support scrollbars. In Microsoft TV, the choices available on each data entry page must fit within the vertical 420-pixel boundary or they will dip below the visible area of the TV screen. Option groups are a good way to offer many choices or lines of information in a minimum amount of space.

Implementing Option Groups

Option groups are created with the <select></select> tags, as shown in the following sample code:

```
<select name="selpizza" style="position:relative; left:5"
onchange="calcprice()">

<option value="gourmet" tabindex=0> Bob's Gourmet Supreme
<option value="veggie"> Garden Veggie Delight
<option value="meat"> Meat Eater
<option value="kahuna"> Big Kahuna Hawaiian

</select>

Quantity

<select name="selquan" onchange="calcprice()">
<option value=1> 1 unit
<option value=2> 2 units
<option value=3> 3 units
<option value=4> 4 units
<option value=5> 5 units
<option value=6> 6 units
<option value=7> 7 units
<option value=8> 8 units
<option value=9> 9 units
</select>
```

Notice that each option has a value attribute. The value attribute of the selected option is sent to the server by the submit method of the <form> tag that contains it. The value of an option may be a number or a string. In this case, string values such as "gourmet" and "veggie" are used to help make the script that uses them more self-documenting. In the Bob's Pizza demo, TV viewers can select from a variety of options, including pizza type, size, and quantity. The calculations necessary to give the user a price, based on the selected options, are performed by the calcprice() function. Notice that in the preceding code sample, the calcprice() function was called by the *onChange* events of both the selPizza and selQuan option groups.

Implementing Radio Buttons

In Bob's Pizza, radio buttons are used to let the viewer select one of three sizes of pizza. Radio buttons present an array of mutually exclusive choices to the TV viewer. In other words, the TV viewer can only select one size of pizza at a time. To create radio buttons, you specify the same name property for each button in the group. For Bob's Pizza, each radio button is given the name of radSize. Also notice in the following code sample that the value of each radio button differs from its mates in the group. The value of the selected button is submitted to the server. The code to implement the radio button group, called radSize, is shown in bold here:

```
Select size

<input type="radio" name="radsize" checked value="12"
onclick="calcprice()"> 12"
<input type="radio" name="radsize" value="14"
onclick="calcprice()"> 14"
<input type="radio" name="radsize" value="16"
onclick="calcprice()"> 16"
```

Implementing Hidden Form Elements

The hidden form element plays a critical role in the Bob's Pizza application. Hidden form elements are used to store data that is not visible to the TV viewer but is submitted with the form.

In the case of Bob's Pizza, the back-end database keeps track of pizzas by type. For example, a 12" veggie pizza is one type, while a 16" veggie is another type. Each pizza type has its own ProductID number and UnitPrice value. When the TV viewer changes values for pizza size or pizza type, the associated ProductID number must be tracked for submission and processing by the back-end database. Because the ProductID has no meaning for the user, its display would only be confusing. The following code shows how the hidden form element is implemented for Bob's Pizza:

```
<!-- Set to product number for default pizza = gourmet 12" -->
<input type="hidden" name="hidprodid" value=4>
```

When Bob's Pizza first loads, the specials.html page opens with Bob's Gourmet Supreme selected in the pizza selection option group and 12" selected by default in the pizza size radio button group. In the back-end database, as shown in Figure 19.5, a 12" gourmet pizza has a ProductID of 4, the same value attribute of the hidden input field named hidProdID in specials.html.

Products : Table		
ProductID	**ProductName**	**UnitPrice**
1	Meat Eater 12"	$12.59
2	Meat Eater 14"	$14.50
3	MeatEater 16"	$16.25
4	Bob's Gourmet Supreme 12"	$15.25
5	Bob's Gourmet Supreme 14"	$17.99
6	Bob's Gourmet Supreme 16"	$19.27
7	Garden Veggie Delight 12"	$12.25
8	Garden Veggie Delight 14"	$14.56
9	Garden Veggie Delight 16"	$17.56
10	Big Kahuna Hawaiian 12"	$13.33
11	Big Kahuna Hawaiian 14"	$16.75
12	Big Kahuna Hawaiian 16"	$19.95

Figure 19.5 *Back-end Access product table for Bob's Pizza.*

NOTE   If you want to adapt this solution for your own use, it is easy to substitute your products in Bob's product table with the same ProductID numbers used here.

The calcprice() Function

Each time a user makes a change to one of the option groups or radio buttons on specials.html, a new price appears in the Price textarea box. This update is accomplished by the calcprice() function, along with the assistance of three helper functions: setSize(), setQuan(), and setType(). The following sample code shows how the calcprice() function is implemented.

```
function calcprice()
{
    var thisform = document.frmPizza;
    var thissize = setSize();
    var thismany = setQuan();
    var thistype = setType();
    var thispizza;

    if (thistype == 'meat')
    {
        thispizza = "Pepperoni, bacon, ham, mushrooms, onions" +
                    " on a double thick dough boy white" +
                    " flour crust";

        thisform.txadesc.value = thispizza

        if (thissize == '12')
        {
            thisform.txaPrice.value = "$" + (12.59 * thismany)
            thisform.hidProdID.value = 1
        }
        else if (thissize == '14')
        {
            thisform.txaPrice.value = "$" + (14.50 * thismany)
            thisform.hidProdID.value = 2
        }
        else if (thissize == '16')
        {
            thisform.txaPrice.value = "$" + (16.25 * thismany)
            thisform.hidProdID.value = 3
        }

        return;
    }
```

```
if (thistype == 'gourmet')
{
    thispizza = "Sausage, olives, mushrooms, peppers, tomatoes," +
                " garlic and pesto on a garlic crust";

    thisform.txadesc.value = thispizza

    if (thissize == '12')
    {
        thisform.txaPrice.value = "$" + (15.25 * thismany)
        thisform.hidProdID.value = 4
    }
    else if (thissize == '14')
    {
        thisform.txaPrice.value = "$" + (17.99 * thismany)
        thisform.hidProdID.value = 5
    }
    else if (thissize == '16')
    {
        thisform.txaPrice.value = "$" + (19.27 * thismany)
        thisform.hidProdID.value = 6
    }

    return;
    }
}
```

The calcprice() function is called by the *onChange* event in the option groups used to select the type and quantity of pizza for each order. The calcprice() function is also called by the *onClick* event of the group of radio buttons used to indicate pizza size. Each time an *onChange* or an *onClick* event happens in one of the option or radio button groups, a new price is calculated and then displayed in the textarea box labeled Price. To achieve an accurate price calculation, the calcprice() function must gather information about the current values for each user choice on the form. To see how this is accomplished, look at the variable initiation statements at the beginning of the calcprice() function:

```
var thissize = setSize();
var thismany = setQuan();
var thistype = setType();
```

Each of these variables—thissize, thismany, and thistype—is filled from the return value of a function specialized in detecting the value of a particular form control. For example, the setSize() function returns the current selection in the group of radio buttons collectively named radSize.

Returning Radio Button Values

Elements in a form that share the same value for their name attribute are placed in an array. The name of the array is the same as the name common to the elements. For example, the radio buttons on specials.html are each called radSize, so the array created for them is *radSize[n]*, where *n* is a zero-based number, one for each radio button. By polling the checked property of each item in the *radSize[n]* array, you can determine which radio button is selected and return its value to the calcprice() function to fill the variable named thissize. Figure 19.6 shows the radio button group named radSize on the specials.html page.

Figure 19.6 *Radio buttons named radSize.*

Here is the code for the setSize function that feeds the variable named thissize:

```
function setSize()
{
    if (document.frmPizza.radSize[0].checked)
    {
        return 12;
    }
    if (document.frmPizza.radSize[1].checked)
    {
        return 14;
    }
    if (document.frmPizza.radSize[2].checked)
    {
        return 16;
    }
}
```

> **NOTE**   An alternate way to return the value of a radio button is to refer to its value. For example, if the 14" button were clicked, instead of returning "14" you could use the following syntax:
>
> ```
> return document.frmPizza.radSize[1].value
> ```

Returning Option Group Values

The variable named thistype in the calcprice() function is filled by the setType() function. This function determines the value selected in the pizza selection option group named selPizza. The setType() function detects which pizza type to calculate a price for, as shown in the following code:

```
function setType()
{
   var pizza;
   thisform = document.frmPizza
   thislist = thisform.selPizza
   thisoption = thislist.selectedIndex;
   pizza = thislist.options[thisoption].value;

   return pizza

}
```

Each option group maintains an options array that begins with zero and goes up to one less than the number of options in the group. The setType() function traverses the object hierarchy to identify the index value of the particular option that the user has selected. This value is returned by the selectedIndex property of the option group. Because returning the index number of the option selected would result in code that is unnecessarily cryptic, the index value is used to return the value of the selected option. Because the values in the selPizza option group are set to readable names such as "meat" and "veggie", the return from setQuan() will be easy to understand. Following is the code for the setQuan() function:

```
function setQuan()
{
   var quantity;
   thisform = document.frmPizza;
   thislist = thisform.selQuan
   thisoption = thislist.selectedIndex;
   quantity = thislist.options[thisoption].value;

   return quantity
}
```

The setQuan() function is nearly identical to the setType() function, so we will not provide a separate explanation of how it works.

How the Pizza Price is Calculated

Once the helper functions setSize(), setType(), and setQuan() have filled the variables thissize, thismany, and thistype with values from the controls on the form, the calcprice() function is ready to do its job.

In the calcprice() function, the variable thistype is tested to see which pizza type the user has selected. If the user chooses "Meat Eater," then thistype will equal meat and the top branch of the "if" statements is executed.

```
if (thistype == 'meat')
{
    thispizza = "Pepperoni, bacon, ham, mushrooms, onions" +
                " on a double thick dough boy"+
                " white flour crust";

    thisform.txadesc.value = thispizza

    if (thissize == '12')
    {
        thisform.txaPrice.value = "$" + (12.59 * thismany)
        thisform.hidProdID.value = 1
    }
    else if (thissize == '14')
    {
        thisform.txaPrice.value = "$" + (14.50 * thismany)
        thisform.hidProdID.value = 2
    }
    else if (thissize == '16')
    {
        thisform.txaPrice.value = "$" + (16.25 * thismany)
        thisform.hidProdID.value = 3
    }
    return;
}
```

The variable thispizza is filled with a string that describes a "Meat Eater" pizza. This value is used to change the text displayed by the textarea control named txadesc, which is located under the pizza selection option control on the form. Note that the description only changes if calcprice() was called by the selPizza option group. If called by the selQuan option group or the radSize radio button group, the value of thispizza does not change. Even when thispizza gets filled and the textarea is changed in the code, the value remains the same, so no change is noted by the viewer.

In the calcprice() function, a series of "if/else" statements are nested within the outer "if" statement to determine the size of the pizza the user has chosen. For example, if the user chooses the radio button labeled 14", the following "else" branch would be entered:

```
else if (thissize == '14')
{
    thisform.txaPrice.value = "$" + (14.50 * thismany)
    thisform.hidProdID.value = 2
```

The textarea box named txaPrice is used to display the order total adjacent to a label that reads "Price." It is changed to reflect the price per unit for a 14" meat pizza, multiplied by the number chosen by the viewer in the Quantity option group. The quantity value is stored in the thismany variable. Finally, the hidden field named hidProdId is set to the ProductID2 (for a 14" meat pizza) as it is listed in the back-end database shown in Figure 19.7.

ProductID	ProductName	UnitPrice	ProductDescID
1	Meat Eater 12"	$12.59	1
2	Meat Eater 14"	$14.50	1
3	Meat Eater 16"	$16.25	1
4	Bob's Gourmet Supreme 12"	$15.25	2
5	Bob's Gourmet Supreme 14"	$17.99	2
6	Bob's Gourmet Supreme 16"	$19.27	2
7	Garden Veggie Delight 12"	$12.25	3
8	Garden Veggie Delight 14"	$14.56	3
9	Garden Veggie Delight 16"	$17.56	3
10	Big Kahuna Hawaiian 12"	$13.33	4
11	Big Kahuna Hawaiian 14"	$16.75	4
12	Big Kahuna Hawaiian 16"	$19.95	4
(AutoNumber)		$0.00	0

Figure 19.7 *Product table in Bob's Access database back end.*

Notice that the ProductID for a 14" meat pizza is 2 and the price is $14.50. In a fully fleshed out production application, these price and ProductID numbers would be dynamically generated rather than being hard-coded. That way, changes made in the database would automatically be reflected on the Web page. You will see more about dynamically bringing content from a back-end database to a Web page in Chapter 20.

> **NOTE**   All the possible combinations of size, type, and price are represented in the "if /if else" decision blocks within the calcprice() function. All of these blocks work in the same way as the example described earlier.

Why Not Use the JavaScript *Switch* Statement?

The seemingly effusive use of "if" statements in the calcprice() function raises the question: Why not use a *switch* statement? No doubt, nested "if" statements are not the most elegant programming structure available. But ECMAScript, outlined by the specification adopted by the ATVEF committee as a content creation standard, does not support the *switch* statement. Therefore, to faithfully follow the ATVEF guidelines, the *switch* statement is avoided in the code samples in this book.

WHAT'S NEXT

In this chapter we described how to create form elements and how to script to them. We also explored several important design principles for creating forms for Microsoft TV. These include the use of the put method in forms that will be submitted to Web servers for use with ASP-ADO pages and e-commerce, and the application of textarea controls to present scriptable read-only text to viewers. Now that we've covered how the order entry form for Bob's Pizza is created, let's look at how to write the form data to an Access database using ASP and ADO technology.

Creating ASP-ADO Code to Interface with the Database

In This Chapter

- Benefits of Using ASP-ADO

- Installing ASP and ADO

- ASP-ADO Basics

- Creating a Shopping Cart Using Order.asp

- Getting Customer Data with the Customer.html Page

- Client-Side Data Validation in Microsoft TV

- Coding ASP for Multiple Tables on Thanks.asp

- Database Maintenance

- Bob's Pizza Database Explained

- Tips and Tricks in ASP-ADO Coding

This chapter introduces the concept of using Active Server Pages (ASP) and Microsoft ActiveX Data Objects (ADO) to create e-commerce functionality for Bob's Pizza, including:

■ How to use ASP and ADO technology to submit data in a form to a back-end Microsoft Access database

■ How shopping cart functionality was implemented in Bob's Pizza

■ How to use ASP and ADO technology to provide dynamic feedback to the TV viewer

After reading this chapter, you should understand ASP and ADO code and relational databases well enough to modify the Bob's Pizza example for your own use.

BENEFITS OF USING ASP-ADO

Active Server Pages (ASP) provide a relatively easy way to process information submitted by HTML forms and dynamically generate return Web pages based on the submitted information. Active Server Pages (ASP), in combination with ActiveX Data Objects (ADO), provide a set of objects that programmers can use to integrate databases with Web pages. ASP objects are used to render data on HTML pages, while ActiveX Data Objects (ADO) are used to transfer data between a Web page and a database. With ASP and ADO technology, developers get high-speed database connectivity, ease of use through a hierarchy-free object model, low memory overhead, and a small disk footprint.

ASP and CGI

The predecessor of and current competitor to ASP pages are Common Gateway Interface (CGI) programs. Believe it or not, every time someone clicks a Submit button on a page that uses CGI, a new instance of the CGI program opens, creates an HTML page, sends the new page back to the client, and then closes itself. If a site gets large amounts of traffic, there can be hundreds of instances of a CGI program opening and closing simultaneously, all using system resources.

One of the main advantages of ASP over CGI is that it runs "in process," meaning that just one instance of the ASP program gets loaded into system memory. And it stays "in process" until the server is rebooted. While in memory, the ASP program uses its Dynamic Link Library (DLL) to create HTML on the fly each time the server receives a request for a page with the .asp extension. This single ASP program may provide better performance than CGI, while simultaneously using fewer system resources. In addition to all these advantages, ASP-ADO is a great deal easier to learn than CGI programming in a language like Perl or C.

The only limiting factor of ASP technology is that it only works if your pages are hosted on a Microsoft Windows NT or Microsoft Windows 2000 server running Internet Information Services (IIS), or on a Microsoft Windows 98 computer running Personal Web Server (PWS).

INSTALLING ASP AND ADO

One of the greatest benefits of ASP and ADO is that the installation is done for you! Both ASP and ADO are automatically installed and registered by a host product such as IIS or PWS. If you installed Option Pack 4 over Microsoft Windows NT 4.0, you have IIS, ADO, and ASP already.

ASP-ADO BASICS

There are thousands of pages written about the use of ADO and ASP. It is well beyond the scope of this book to cover these topics in detail. When reading this chapter, it is important to understand that ADO and ASP provide many ways to solve a problem. In building Bob's Pizza, choices had to be made between writing code that would be easy to understand and copy, or code that would generate the least server load. In most cases, easy-to-understand code was chosen over code that would provide optimal speed and efficiency. For example, all the SQL strings used in this chapter may be built using the Access Query by Example (QBE) grid. More advanced SQL statements like *Insert*, *Update*, and *Delete*, which are difficult or impossible to create in the QBE grid, were not used.

The methods shown in this chapter will work fine on a regional application like Bob's Pizza. However, if you plan to create a central site with a national or global scope, it would be a good idea to consult a SQL, ADO, and ASP expert. The expert will be able to leverage the server clustering capability of Microsoft TV Server to make sure your site will scale to the degree necessary to handle tens of thousands of simultaneous hits.

ASP Objects and Server-Side Scripts

At its core, ASP technology is easy to understand. When a Windows NT or Windows 2000 server running IIS or a Windows 98 machine running PWS receives a request for a file with an .asp extension, it knows that the page will be an ASP page. For example, order.asp is a page that is generated on the fly by ASP for Bob's Pizza. ASP scripting is always contained in a funny-looking set of brackets such as the following:

```
<%asp server side script%>
```

Unless otherwise informed, Windows Web server software runs script that occurs within the <% %> brackets on the server side. Generally, ASP script is used to generate useful information on the client's browser. For example, the following "Hello" code might appear on a page that informs TV viewers of the current time and Bob's Pizza's hours of business:

```
<h2>Welcome to Bob's Pizza</h2><br><br>
<h3>It is <%= time()%><br><br>
Bob's Pizza is open 9AM to 5PM<br>
Monday through Friday
```

In the preceding code sample, notice the <%= sign in the set of ASP brackets. ASP brackets that contain an equal sign (<%= myvariable %>) are used to draw variable values or the result of functions. Plain ASP tag sets (<% script stuff %>) contain script that runs on the server side, but is not necessarily shown on the client browser. ASP code is browser-agnostic because it generates plain HTML, as shown in Figure 20.1, which shows how the "Hello" code renders in the Microsoft WebTV Viewer.

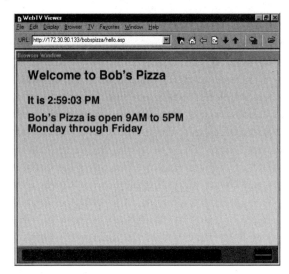

Figure 20.1 *The "Hello" code rendered in the WebTV Viewer.*

Defining the Server-Side Scripting Language

You will find server-side scripting on each ASP page in Bob's Pizza. For example, the following script tells IIS that the ASP server-side script on the order.asp page is written in the Microsoft Visual Basic Scripting Edition (VBScript):

```
<% language = vbscript %>
```

> **NOTE**   VBScript is the default language interpreted by the ASP engine and does not have to be explicitly set. If you use JavaScript instead, you have to explicitly set the language as follows: <% Language = JScript %>.

ASP Objects Used in Bob's Pizza

Bob's Pizza uses the following ASP objects:

- **Request object**—Gets information from the TV viewer. You can get HTML form data, read cookies, and so forth.

- **Response object**—Sends information to the TV viewer. You can send text to the page, redirect to another URL, and set cookies.

- **Server object**—Interacts with the server. You can access a database, read files, and find out about the capabilities of the browser.

- **Session object**—Enables you to manage information about the current session. (Each TV viewer has one session per open browser.)

The following sections describe in more detail how the objects are used.

> **NOTE** For more information about other objects that come with IIS and ASP, see *http://msdn.microsoft.com/workshop/server/asp/aspfeat.asp?RLD=22.*

ADO Object Model

By now, some of you may be saying: "Oh no, an object model!" Well relax, the ADO object model is easy compared to some of the hierarchical models such as Data Access Objects (DAO) and Remote Data Objects (RDO) which you may have seen in the past. Besides, this section doesn't beat this subject to death. Instead, it presents only enough information to allow you to make sense out of the rest of this chapter.

There are only three main objects in the ADO object model used in Bob's Pizza:

- **Recordset object**—provides the main interface to data.

- **Command object**—is particularly useful in squeezing those last few nanoseconds of performance out of a database management system.

- **Connection object**—represents open connections to back-end database systems.

ADO Recordset Object

Recordset objects represent a set of records in column and row format. Each row typically contains several fields of *varied* type about a particular thing. Each field represents part of a column of data of *like* type. For example, in the customer table of Bob's database, rows of data contain fields named CustomerID, name, address, and phone number. Each row has information about a particular customer, made up of fields with data of various types. For example, all the CustomerID fields hold data of the type *long* and all the name fields hold *string* type data.

Recordsets come from underlying tables or as the result of a query run on the tables in a back-end database. The Open method of a Recordset object is used to open a recordset. Parameters of the Recordset object Open method are shown here:

```
RS.Open Source,ActiveConnection,CursorType,LockType,Options
```

Parameters are used to set properties of a recordset retrieved by the Open method. For example, the following code represents the minimum VBScript needed to open a recordset containing the entire customer information in Bob's database Customer table:

```
Set myRs = Server.CreateObject("ADODB.Recordset")
    myRs.Open "Select * From Customer", "database=Bobs;DSN=Bobs"
```

In the preceding example code, a Recordset object is created, and then its Open method is given a SQL string as the value of its *Source* parameter. The SQL string shown may be read as "Select all (*) the records from the Customer table." The *ActiveConnection* parameter is filled with a Data Source Name (DSN). The database named by the DSN Bobs contains the Customer table.

The recordset held in the variable myRs is a default forward-only read-only recordset. This means all you can do is iterate through it in a top-down manner and maybe render it on screen. ASP techniques used to render records on screen will be explained momentarily. If a more functional recordset is required, say to write data back to the database from the Web page, more parameters of the Open method need to be set. There is a much more complete discussion of the Recordset object and the parameters of its Open method coming up. For now, take a look at the following code:

```
Set myRs = Server.CreateObject("ADODB.Recordset")
    myRs.Open "Select * From Customer", "database=Bobs;DSN=Bobs",
    adOpenKeyset, adLockBatchOptimistic
```

By setting two more parameters of the Recordset Open method, a recordset is created that can be written to, as well as read. Also, moving backward, as well as moving forward, is supported, which is important to the Bob's Pizza application, as we'll soon discuss.

ADO Connection Object

The ADO Connection object represents open connections to back-end database systems. In the section on the ADO Recordset object, you saw that it is possible to connect to a database and retrieve records using only the Recordset object. This may make you wonder why a Connection object is needed at all.

The answer is that Connection objects can improve the efficiency of a page and reduce server load when a page uses multiple Recordset objects. Rather than establishing a connection for each recordset, the ActiveConnection property of a Recordset object may be set to a variable representing a Connection object. For example, the following code opens recordsets on the Products and Orders tables in Bob's database using a single Connection object:

```
Set myConn = Server.CreateObject("ADODB.Connection")

Set myRsProducts = Server.CreateObject("ADODB.Recordset")
Set myRsOrders = Server.CreateObject("ADODB.Recordset")

myConn.Open "Bobs"
myRsProducts.ActiveConnection = myConn
myRsProducts.Open "Select * From Products"
'do something like use asp to display product data

myRsOrders.ActiveConnection = myConn
myRsOrders.Open "Select * From Orders"
```

The connection information held in the Connection object is read by the server much faster than the raw connection string used in the earlier example. The trade-off for this efficiency is server memory. Every object created uses memory until it, or the page that contains it, is closed.

The Connection object exposes a handy method named Execute that may be used to return Recordset object references without ever creating a Recordset object! For example, the following code returns a read-only recordset containing all the customer data in Bob's database using only a Connection object and the Execute method:

```
Set myConn = Server.CreateObject("ADODB.Connection")
myConn.Open "Bobs"
Set myConnRs = myConn.Execute("Select * From Customer",,adCmdText)
```

Using the Execute method of the ADO Connection object is one way to reduce server load when forward-only read-only recordsets will do the job. As previously stated, more functional recordsets are returned by the Open method of the Recordset object.

ADO Command Object

The ADO Command object is called "optional" because not all database management systems support its methods. If your database management system (DBMS) supports them and parameter queries are part of your world, its worthwhile to investigate the Command object. For the rest of us, it is enough to know that a parameter query is used when values are required to get an answer from your data source and the Command object is the best way to pass those parameters. For example, you may have a built-in query that returns sales data if you give it a beginning and ending date. If your application uses this query multiple times, server load can be ameliorated and performance increased by passing date parameters using the Parameters collection of the Command object. We do not use the Command object in Bob's Pizza, so this is the last you will hear about it in this book.

ADO Constants

Many of the ADO code examples you have seen so far contain constant values used to set parameters in object methods. For example, the Open method of the ADO Recordset object employed two constants, *adOpenKeyset* and *adLockBatchOptimistic,* that modified the *LockType* parameter. In Windows programming languages such as Microsoft Visual Basic where your code runs on the desktop, you can assume that the compiler knows what constants like *adOpenKeyset* and *adLockBatchOptimistic* mean. In the disconnected world of client-server Web programming, there is an extra step you must use in order to employ constants in your code.

Using Constants

At the top of every ASP page in Bob's Pizza there is a Server-Side Includes (SSI) statement that looks like the following:

```
<!-- #Include file="adovbs.inc" -->
```

The file adovbs.inc contains all the VBScript constants used with ASP pages. This file needs to be on your server in the same folder where the Bob's Pizza pages are located.

> **NOTE**   If you are using JavaScript for your server-side code instead of VBScript, you must include the adojavas.inc file instead of the adovbs.inc.

Many ADO objects use constants to set their properties. Using constants instead of numeric equivalents makes your code more self-documenting. The downside of using constants is that they leave your code open to the slings and arrows of outrageous corruption or loss of the SSI file.

CREATING A SHOPPING CART USING ORDER.ASP

One of the most popular features of modern e-commerce application is a "shopping cart," where TV viewers can accumulate products before submitting the entire order. In Bob's Pizza, the shopping cart is under the control of the Shopcart table and the order.asp page. This section shows how ADO and ASP code is used to write data to the Shopcart table of the Access database that holds Bob's Pizza data. This section also shows how HTML is combined with ASP-ADO to display any single TV viewer's total current order.

Viewing the Order.asp Page

The order.asp page is the page that TV viewers see after they have submitted an order for a pizza. The order.asp page collects the orders made by the TV viewer and enables them to add or remove pizzas from the shopping cart.

To view the order.asp page:

1. Open the WebTV Viewer.

2. From the File menu, click Open File.

3. Browse to the BobsPizza folder and click on base.html.

4. Select a pizza, and then click the Order button to open the order.asp page.

5. Click the Add button and order another pizza.

6. Repeat the order sequence until there are at least three pizza orders in the Order box, as shown in Figure 20.2.

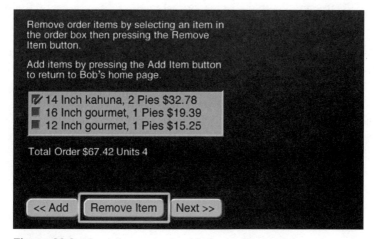

Figure 20.2 *The order.asp page with three pizza orders.*

Shopping Cart Overview

There are two pages that control the shopping cart behavior in Bob's Pizza:

■ specials.html, where order items are created

■ order.asp, where orders are accumulated

The following steps provide an overview of how specials.html, order.asp, and a database work together to create the shopping cart behavior for Bob's Pizza:

1. An order such as two 14-inch kahuna pizzas is created on the specials.html page, and then the Order button is pressed.

2. The form on specials.html containing information about the order is submitted to the Web server using the Post method. The form Action is set to send order.asp back from the server, as shown in the following code snippet:

```
<form name="frmpizza" method="post" action="order.asp">
```

3. On the server, a special collection named Form of the ASP object named Request holds information submitted from the form. Each input control on the form is an item in the Request.Form ("items") collection. For example, there is an option group on the specials.html page with its name attribute set to selPizza. The selected pizza type from the selPizza option group is submitted with the other form elements and ends up on the server as: Request.Form("selPizza").

4. ActiveX Data Objects (ADO) are used with scripting code to open a recordset on a table named Shopcart in the back-end database. Each record in the Shopcart table represents an item in the shopping cart made up of a particular pizza type, size, and quantity. Each field name in the table is held in the Fields collection of the Recordset object. For example, the following line of code sets the field named PizzaType to the value passed by the form element named *selPizza*:

```
ObjRs.Fields("PizzaType") = Request.Form("selPizza")
```

   Figure 20.3 shows the field structure of the Access table named Shopcart.

5. The ADO recordset that writes data to the Shopcart table is closed, and a faster recordset is opened containing rows of data about the particular TV viewer. All the orders the TV viewer has assembled on specials.html are sent to order.asp and appear in the recordset rows.

6. This TV viewer's combined order recordset field data is combined with HTML tags to write the order information displayed on the order.asp page, as shown in the following code snippet:

```
<option label="gourmet" value="<%= ordercount %>">
<% = objRs("PizzaSize")%> Inch <%= objRs("PizzaType") %> --
<%= objRs("PizzaQuan")%> Pies For <%= objRs("PizzaTotal")%>
</option>
```

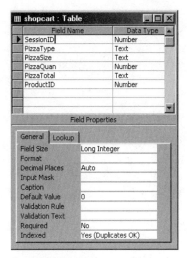

Figure 20.3 *The Shopcart table in an Access database on the server.*

For example, by iterating through a three-row recordset the following HTML and recordset field data was added to the option group on order.asp, as seen in Figure 20.4.

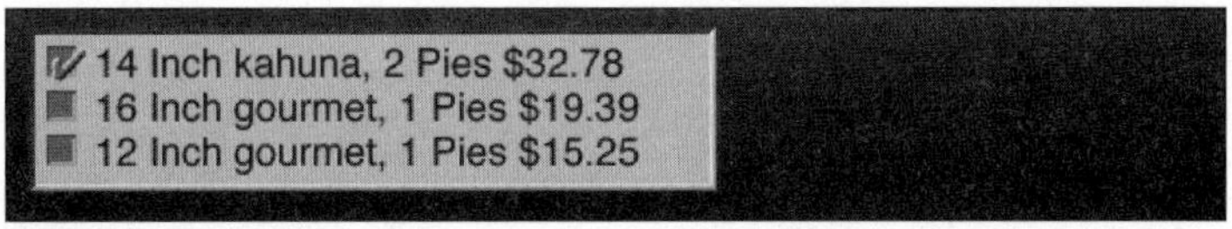

Figure 20.4 *Order box on order.asp with three records.*

7. If the TV viewer returns to the specials.html page from the order.asp page by using the Add button and he picks another pizza, it is added to the shopping cart. Since previous order items are held in the table named Shopcart, they are displayed on order.asp along with the new order, as seen in Figure 20.5. Notice the small arrow in the lower right of the Order box, indicating that scrolling is possible.

Figure 20.5 *More than three rows in the order box cause a scroll arrow to appear.*

There is a reason that ASP is often referred to as server-side script. You see, all the reading and writing of data happens on the server side in a split second. When order.asp is called, the Shopcart table is opened and a new row is written in it. This record includes a number unique to the TV viewer and the pizza size, type, and quantity values passed by the form elements on specials.html to the ASP-ADO code on order.asp. After the new order is written, the first recordset is closed. A new recordset opens containing just the TV viewer's rows from the Shopcart table. This new leaner recordset is used with ASP to render HTML so the TV viewer's choices may be seen on screen.

Removing a Record from the Shopcart Table

The user may remove items from the Shopcart table, and hence from the shopping cart, by using the Remove button on order.asp. This is accomplished by sending the record number of the item to remove to the server, using a hidden input that is part of a <form> on the order.asp page. The method of this <form> is set to send the page back to itself, as shown in the following code:

```
<form name="setorder" method = "post" action="order.asp" >
```

An ADO recordset is opened on the Shopcart table, containing only records for this TV viewer. The designated item is removed, and order.asp is rendered with one fewer item in the Order box. The following code shows how this is accomplished:

```
<%
    If objRs.BOF and objRs.EOF Then

        ordertotal = "None"
        unitcount = "None"

%>
<option> No Pizzas Currently Ordered </option>
<%

    End If
%>
```

If all the items are removed, a test for the empty recordset is met and the Remove button is disabled, as shown in Figure 20.6.

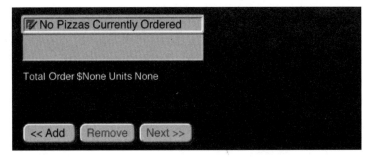

Figure 20.6 *The order.asp page with the Remove button disabled.*

Do not be concerned if you feel confused about how the shopping cart works. The rest of this section is devoted to fleshing out the concepts presented in the shopping cart overview.

Connecting to a Database and Opening It

Connecting to a database is relatively easy. It involves using the built-in ASP Server object to create an instance of the Connection object, and then using its Open method, as shown in the following code:

```
Set objConn = Server.CreateObject("ADODB.Connection")
```

> **NOTE** In the preceding code example, notice that creating object variables requires use of the Visual Basic Scripting Edition (VBScript) *Set* statement.

With the Connection object established, the next step is to use it to open the correct database on the server. The parameters of the Connection object's Open method are shown below.

```
objConn.Open ConnectionString, UserID, Password
```

For Bob's Pizza, the TV *UserID* and *Password* parameters are not required. Only the *ConnectionString* parameter of the Open method needs to receive a value. The *ConnectionString* parameter of the Connection object's Open method uses the database's DSN to do its job. For example, the DSN for the database used in Bob's Pizza is "Bobs".

```
'Open connection to DSN named Bobs on server
Set objConn = Server.CreateObject("ADODB.Connection")
objConn.Open "Bobs"
```

> **NOTE** If you missed how to set up a DSN, see Chapter 19, "Building Bob's Order Entry Page." If you plan to work with database management products such as Microsoft SQL Server, you must provide TV viewer and password information as parameters to the Open method. Your database server administrator would be a good person to ask about connection string values such as *UserID* and *Password*.

Creating a Recordset Object

After connecting to the Bob's Pizza database and opening it, the next step is to create a Recordset object. ADO recordsets are the winged workhorses that bring values from the back-end database (server) to the front-end browser (client). Think of ADO recordsets as rows and columns of information that can be transmitted across a local area network (LAN) connection or the Internet.

To create the Recordset object, the ASP server object's CreateObject method is used again. This time a Recordset object, rather than a Connection object, is created:

```
Set objRs = Server.CreateObject("ADODB.Recordset")
```

After the Recordset object is created, its Open method is used to make a set of records available for use. To create a set of records, the Open method must know the database to write to and the table, or parts of tables, to use. Several properties of the new Recordset object must be set before the recordset is opened. These properties determine the type of recordset that is created. The following code snippet shows the code used to set the properties of a Recordset object:

```
'Set properties of recordset object and open it on table named shopcart
'setting Options parameter in Open method helps optimize speed

        Set objRs = Server.CreateObject("ADODB.Recordset")
        objRs.ActiveConnection = objConn
        objRs.CursorType = adOpenKeyset
        objRs.LockType = adLockOptimistic
        objRs.Source = "shopcart"
```

The fastest recordsets have specialized uses, such as the simple display of data, while more complex recordsets are used to change existing data or write new records to the underlying tables. The parameters used to shape a Recordset object include *ActiveConnection*, *CursorType*, *LockType*, and *Source*, as shown in the following ADO Recordset Open method syntax:

```
objRst.Open Source, ActiveConnection, CursorType, LockType, Options
```

> **NOTE**   You may be wondering why "ADODB" is placed in front of each ADO object created. ADO is a dual interface COM type library. In two- and three-tiered database applications, the Program ID (ProgID) is ADODB. In contrast, ADO code used on single-tier applications executed on the client use a ProgID of "ADOR".

Setting the *ActiveConnection* Parameter

Setting the *ActiveConnection* parameter to a variable that represents a Connection object will point the Recordset Open method to the appropriate database. In the case of Bob's Pizza, the objConn variable holds the Connection object that opens the "Bobs" database:

```
objRs.ActiveConnection = objConn
```

Setting the *CursorType* Parameter

The *CursorType* parameter determines how functional the recordset will be. The default value is *adOpenForwardOnly*. The best thing about the default is it's fast. In situations where all you want to do is display the data on a Web page, the *adOpenForwardOnly* cursor is a champ. Microsoft Access, however, needs a more functional cursor in order to update a database table. The most capable cursor that works on Microsoft Jet (the engine underlying Access) is the *adOpenKeyset* cursor, as shown in the following code sample:

```
objRs.CursorType = adOpenKeyset
```

The *adOpenKeyset* cursor makes it possible to move forward and backward in the recordset, and to change or add data to the table it is based on.

Setting the *LockType* Parameter

The *LockType* parameter is used to prevent data collisions. A data collision occurs when two or more TV viewers try to write data to the same record at the same moment. The default setting is *adLockReadOnly*. As its name implies, the default setting only allows the data to be read. Just as with the default cursor, the default *LockType* setting is built for speed, not for comfort. Leaving the *LockType* property with its default value produces a very fast recordset, but all you can do is look at it. Because the recordset used in Bob's Pizza writes data to the underlying table, a more functional *LockType* is needed. Thus, the *LockType* is set to *adLockOptimistic*, as shown in the following code:

```
objRs.LockType = adLockOptimistic
```

Optimistic locking freezes the recordset for the fraction of a second it takes for the Update method to update the underlying table with new data. You can contrast optimistic locking to pessimistic locking. Pessimistic locking (*adLockPessismistic*) freezes the recordset from the moment it is opened to the moment it closes or goes out of scope, such as when the page closes. Pessimistic locking reduces the chance of a collision, but increases the chance a TV viewer will get an "Access Denied" error. Optimistic locking is the best compromise to use for Bob's Pizza.

Setting the *Source* Parameter

An earlier section in this chapter demonstrated how the *ActiveConnection* parameter points to a particular database. The *Source* parameter defines the particular data within that database that the recordset should open. In this case, the recordset opens the Shopcart table. As shown in the following code, the recordset defined by the variable objRs is set to open the Shopcart table:

```
objRs.Source = "shopcart"
```

Optimizing the Recordset with the *Option* Parameter

As far as ADO knows, you might be trying to open a table or a stored procedure (query), or you might have passed a SQL statement to the *Source* parameter. By telling ADO the kind of recordset you are opening, you can optimize performance.

To define the type of recordset, you use the *Option* parameter of the recordset. This property tells ADO the type of table or query represented by the Source property. For this part of Bob's Pizza, the *Option* parameter is set to the *adCmdTable* constant. This constant tells ADO to open a recordset on a table in the underlying database. The two values used in this demo for the *Option* parameter are:

- *adCmdText* is used with SQL strings as the Source.
- *adCmdTable* is used with tables as the Source.

Opening the Recordset Object

Now that a connection to Bob's database has been established and all the properties of an updateable recordset have been set, it is time to actually open a recordset. The following code shows the Open method used to open a Recordset object:

```
objRs.Open , , , ,adCmdTable
```

Notice that the Open statement uses commas for the *Source*, *ActiveConnection*, *CursorType*, and *LockType* properties. Since these properties have already been set, you do not have to set them again.

Adding a Record Using the AddNew and Update Methods

Now that the recordset is open, something must be done with it! That something is adding a new record. The new record consists of data passed by the form elements on specials.html to order.asp.

To add a new record to the underlying database, you use the AddNew and Update methods of the Recordset object. Syntax for the AddNew method is very simple, as shown in the following code:

```
Recordset.AddNew Fields, Values
```

Fields is the default collection of a Recordset object. This collection includes all the fields defined for the table opened by the Recordset object. For this example, these fields represent cells in the Shopcart table. This combination of fields gets passed to the AddNew method and comprises one record (row) in the Shopcart table.

Values are the data values you want placed in the fields. These values come from the selections made on the form and are held on the server in the *Form* collection of the ASP Request object. Request is the second ASP object use thus far. First, the CreateObject method of the ASP Server object was used to create ADO Connection and Recordset objects. Now, the *Form* collection of the ASP Request object is going to be used to provide values to the AddNew method of the ADO Recordset object. It's like hammer and tong the way ADO and ASP work together!

The Request Object Form Collection

Using Bob's Pizza as an example, the viewer selects a pizza type on the form defined by specials.html. This selection gets assigned to *selPizza* and placed on the server as *Request.Form("selPizza")*. The following code shows the AddNew and Update methods of an ADO Recordset object:

```
objRs.AddNew
    objRs("PizzaType") = Request.Form("selPizza")
    objRs("PizzaSize") = Request.Form("radSize")
    objRs("PizzaQuan") = Request.Form("selQuan")
    objRs("PizzaTotal") = Request.Form("txaPrice")
    objRs("SessionID") = Session.SessionID
    objRs("ProductID") = Request.Form("hidProdID")
objRs.Update
```

Notice each entry in the *Form* collection gets assigned to its corresponding field in the Shopcart table. For instance, the *selPizza* entry gets assigned to the PizzaType field. Also notice that, after all the values have been assigned for each field, the Update method is called to commit these values to the database. This is where the locking takes place. Because we set the *LockType* parameter to optimistic, the underlying table is locked for the split second it takes to write the data. This concept of disconnected recordsets is key to your understanding of client-server programming. The server opens a certain type of recordset, then completely forgets who opened it. Hundreds of other people could be opening similar recordsets at nearly the same instant. When the Update method is reached, the server remembers who opened it and gives him exclusive access to the table for the nanosecond needed to accomplish the task.

Tracking Orders Using the SessionID Property

There is one field in the Shopcart table that is not filled by the Request.Form collection. That field is SessionID, as shown in the following code sample:

```
objRs("SessionID") = Session.SessionID
```

The SessionID field must contain a value that is unique for every TV viewer of Bob's Pizza. Without a unique ID number, there would be no way to tell one TV viewer's order from another.

Luckily, ASP provides a Session object that is just great at providing ID numbers. ASP creates a Session object for each client requesting a page from Bob's Pizza. The Session object has a SessionID property that returns a long number. For example, Figure 20.7 shows how the SessionID field of the Shopcart table contains order information for two TV viewers, 72719533 and 72719534.

shopcart : Table

	SessionID	PizzaType	PizzaSize	PizzaQuan	PizzaTotal	ProductID
▶	72719533	veggie	16	2	$34.78	9
	72719533	kahuna	12	1	$13.39	10
	72719534	kahuna	14	2	$32.78	11
	72719534	gourmet	16	1	$19.39	6
	72719534	gourmet	12	1	$15.25	4
*	0			0		0

Record: 1 of 5

Figure 20.7 *The Shopcart table with data from two TV viewers.*

As shown in Figure 20.7, the Shopcart table has two customers. The customer with the SessionID number of 72719533 is ordering two veggie pizzas and one kahuna. The customer with the SessionID of 72719534 has one 16-inch gourmet pizza on deck, one 12-inch. He also has two 14-inch kahunas—hungry boy.

NOTE SessionID values are stored on the server for 20 minutes or until the client's browser closes.

Maintaining State with Session

One of the amazing things about the SessionID property of the ASP Session object is that it maintains state across pages. This is how the Bob's pizza application manages to remember what each customer wants when he or she goes back and forth between the pages. It even works if the TV viewer changes television channels or otherwise leaves Bob's site and comes back. For example, suppose in the middle of ordering some pizzas that the TV viewer gets an uncontrollable urge to find out the wombat population in Queensland, Australia. He can search and surf to his heart's content (or 20 minutes when the ID times out), and when he returns to Bob's Pizza, his order will be just as he left it.

Closing the Recordset Object

As amazingly efficient as ADO recordsets are, they do put a load on the server. Just as with the light in the bathroom, you should turn off an ADO Recordset object when you are done with it. After a record has been entered, you use the Close method to close the recordset held in the objRs variable that represents the Recordset object. Just closing a recordset does not actually free up the system resources it is using. You need to actually remove the Recordset object from memory to free up the resources. To do that, you use must send the variable objRS spinning off into the void by using the *Set* statement and the keyword *Nothing*:

```
objRs.Close
Set objRs = Nothing
```

If a page has finished processing, the variables that hold them go out of scope, and the steps of closing a Recordset object and removing it from memory are handled automatically by ASP. In spite of this automatic feature, it is good coding practice to remove objects from memory when your code is done with them.

> **NOTE** Using recordsets and cursors is a conceptually easy way of writing data to a database table. It may not be the optimal way for high traffic sites. See any SQL manual on using temporary tables, the *Insert, Insert Values, Insert Select* and *Delete* statements on performing complex optimized data manipulation.

Dynamically Rendering HTML Pages

Up to now, this chapter has addressed writing data to the back-end database. Some ASP technology has been employed, but for the most part, ADO has been the star of the show. This section shows you how ADO and ASP can team up to perform one of the coolest feats in Web design, dynamically rendering a page based on input from the TV viewer and a back-end data store.

Introducing the "Racing" Recordset

To write data in a table, we used what could be termed a "Cadillac" cursor. It was a fully functional recordset, built for comfort and had all the optional features. The price of all that functionality was speed and system resources. However, to render data on the order.asp page, a stripped-down cursorless recordset is used. You can read data from this fastest of all recordset types, and it uses the least server system resources. Rendering HTML and data on the screen is what this "racing" recordset is best at. To create this forward-only cursorless recordset, the Execute method of the Connection object is used, as shown in the code on page 310.

```
<%
If Not IsEmpty(Request.Form) Then
   Dim ordertotal
   Dim ordercount
   Dim unitcount
   Dim sqlString
   Dim thisSession

   thisSession = Session.SessionID
   sqlString = "Select * From shopcart where SessionID =" _
      & thisSession
```

In the preceding code, notice the variable named sqlString. It contains a SQL statement that makes sense to DBMS systems. In fact, SQL is sort of the *lingua franca* of database languages. The same string of text, with minor variations, works on Access, Microsoft SQL Server, Sybase, or Oracle database systems. The SQL string held by the variable sqlString translates to the following: "Please give me a recordset that contains all the rows in your Shopcart table, where the field named SessionID is equal to my friend's ID held in the variable thisSession." Since the variable thisSession holds the SessionID for the TV viewer looking at the page, only his pizza selections will be returned for display on the page.

If you are thinking "Yeah, but how do I figure out a SQL string for my project?" or "What do you mean, it brings in a recordset with only this TV viewer's data?", fear not, in the section of this chapter on tips and tricks you will see how to make Access do your SQL talking for you and an explanation of how queries filter data.

> **NOTE**   Many of the design decisions in Bob's Pizza were made with the idea that you will want to modify the code for your own use. For example, the SQL strings used in Bob's Pizza can all be created using the Access QBE grid. For more information about how to create a SQL string using Access, see the "Tips and Tricks in ASP-ADO Coding" section later in this chapter.

The Execute Method of the Connection Object

To get cheap, fast, cursor-free data, we will open a recordset in a different way, using the Execute method of the ADO Connection object. To see this in your text editor, scroll down just below where the Recordset variable objRs was set to Nothing and look for the following line of code:

```
Set objRs = objConn.Execute(sqlString,,adCmdText)
```

Notice that the variable name objRs was recycled, but it bears no relation to the Recordset object that was set to Nothing. The Set statement means objRS is born anew. The variable named objRs is filled with rows from the Shopcart table using the Execute method of the Connection object held in the variable objConn, set earlier. Syntax for the Execute method is as follows:

```
Set recordset = connection.Execute(CommandText, RecordsAffected, Options)
```

> **NOTE**   Incidentally, if you are executing a SQL statement that is not intended to return rows, you may leave the () brackets off the Execute method. Examples of a SQL string command text that need not return rows are Insert and Delete. See SQL documentation for more detail.

Just as when we opened the "Cadillac" record set, we want to ease the strain on ADO by telling it up front what kind of recordset we are asking for. In this case the value passed to the *CommandText* parameter of the Execute method is a SQL string. When the Execute method of the Connection object is using a SQL string to open a recordset, the constant *adCmdText* optimizes performance. Notice that no properties of the Recordset object were required. In the next section of this chapter, you will see how this sleek and speedy forward-only read-only recordset is used to render information on screen.

Dynamically Rendering the Option Group

When ASP is teamed with ADO, Web pages may be rendered dynamically with database driven content. For example, on the order.asp page, the option group labeled Order Box is designed to display the custom pizza order for the current TV viewer. The contents of this shopping cart option group must be dynamically rendered. The <form> and opening <select> tags are static. They do not change with data returned by the recordset, so they are written in standard HTML syntax, as shown in the following code:

```
<form name="setorder" method=post action="order.asp">

<!-- Option group named selorderitem is filled by looping
    through a recordset. -->

<select name="selorderitem" size=3  autoactivate>
```

Notice the AutoActivate attribute of the selOrderItem Select Option Group. The AutoActivate attribute is proprietary to Microsoft TV. Without AutoActivate turned on, the select group is selected, rather than a particular option, when the television viewer moves the selection box to the Select Option Group. Using AutoActivate in Microsoft TV makes the Select Option Group behave like its Windows counterpart. ASP <% %> brackets enclose code that steps through the recordset:

```
<% Do While Not objRs.EOF

    ordertotal = ordertotal + CCur(objRs("PizzaTotal"))
    ordercount = ordercount + 1
    unitcount = unitcount + objRs("PizzaQuan")

%>
```

In the preceding code, notice that the code within the <% %> ASP brackets begins a Do loop that runs until End of Field (EOF) test is met. The EOF in a recordset is one below the last row in the table of data returned. The deep thinkers among you may be wondering, "So, how can we loop in a forward-only recordset?" The answer is, "We do not loop." Instead, this Do loop acts just like a For Next structure. In this way, it is easier to test for the end of the recordset. Also, the empty recordset (EOF AND BOF) has a special use that you will soon see.

Each record adds its dollar value to the ordertotal variable, increments the ordercount variable by one, and adds the number of pies to the unitcount variable. The ordertotal variable stores a dollar total, and the unitcount variable stores the number of pies in the shopping cart.

The ASP code can be broken up with standard HTML tags. For example, the Do loop starts with an opening ASP script tag "<%", then the closing "%>" ASP script tag is used, followed by HTML code used to write an option into the group for each record, as shown in the following code:

```
unitcount = unitcount + objRs("PizzaQuan")
%>
<option label="gourmet" value="<%= ordercount %>">
<% = objRs("PizzaSize")%> Inch <%= objRs("PizzaType") %>
-- <%= objRs("PizzaQuan")%> Pies <%= objRs("PizzaTotal")%>
</option>
```

One row in the recordset writes one option in the Order box. Notice how words are interspersed with values from the recordset to form each option. If the TV viewer had three rows in his shopping cart record, three options would appear in his Order box. Looking at the first row in Figure 20.8, the numeral 14 came from <%= objRs("PizzaSize")%>, the word "Inch" was literal, the word "kahuna" and the quantity value came from <%= objRs("PizzaType") %> and -- <%= objRs("PizzaQuan")%>, respectively. The word "Pies" is literal and the value $32.78 came from <%= objRs("PizzaTotal")%>.

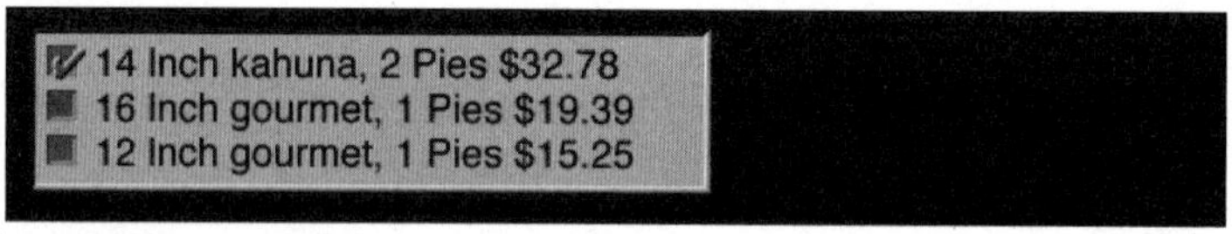

Figure 20.8 *Order box with three orders.*

Each TV viewer sees a page dynamically rendered to reflect his order. Each time a row is written in the shopping cart option group, the following ASP code block is entered and another row in the recordset is accessed:

```
<%
objRs.MoveNext
Loop

%>
```

Rendering the Totals Table

The HTML table that displays values for the total shopping cart order, as shown in Figure 20.9, is rendered using the variables ordertotal and unitcount interspersed with standard HTML and literal words.

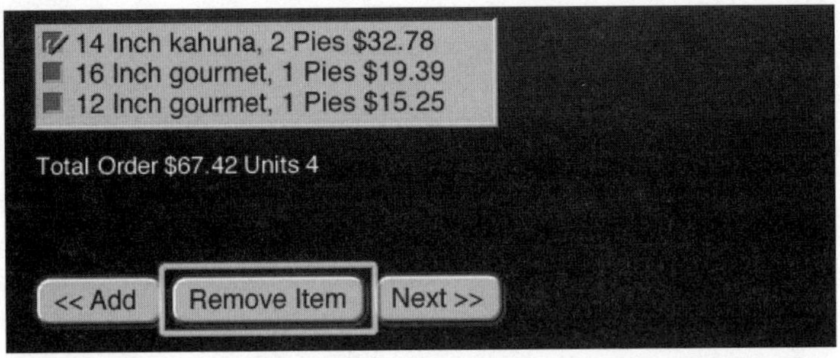

Figure 20.9 *The Totals table displays shopping cart totals with ASP and HTML.*

The following shows the code used to create the Totals table:

```
<table border=1 class="description">
   <tr><td> total order </td><td>$<%= ordertotal%></td>
   <td>units </td><td><%= unitcount%></td></tr>
</table>
```

Deleting Data on the Database

Removing records from a back-end database is a bit harder than adding new records. The reason it is difficult to remove records is that the TV viewer usually wants you to remove some particular record. To remove an item in Bob's Pizza, the TV viewer selects a record in the shopping cart option group and then clicks the Remove button. This submits the form on order.asp to the server. The server opens a recordset containing only records from this TV viewer, moves to the selected record, deletes it, and returns a new copy of order.asp without the selected item. The hard part is finding a way to tell the server how many moves to make in the recordset in order to find the record that the TV viewer wants to delete. To see how this is accomplished, open order.asp in a text editor and scroll down to code that resembles the following:

```
'If no delete requests exist add a new record

If IsEmpty(Request.Form("selOrderItem")) Then

   'AddNew method code here

Else
   'Move to selected record and Delete
   'Use SQL string with SessionID
   'So only this customer's records are affected
```

(continued)

```
objRs.Open sqlString, ,adCmdText
Dim Moves
Moves = Request.Form("selOrderItem") - 1
objRs.Move Moves
objRs.Delete
```

```
End If
```

There are two ways that the ASP script may execute on the order.asp page. If the TV viewer clicks the Order button on the specials.html page, the form submits data, opens order.asp, and writes data using the AddNew method in the top of the If Else, as shown in the preceding code. The other way that order.asp is opened is if the TV viewer clicks on the Remove button on order.asp. This submits the value of the Select Option Group named selOrderItem on order.asp. Counting from the top, the value passed by selOrderItem (the selected item) is 1 greater than the number of moves required to get to the record the TV viewer wishes to delete, as shown in the following code sample:

```
If IsEmpty(Request.Form("selOrderItem")) Then

    'AddNew method code here
```

```
Else
    'Remove item code here
```

If the field Request.Form("selOrderItem") contains a value (is not empty), it indicates the TV viewer has selected an item to delete and pressed the Remove button, so the Else branch is entered. Just a reminder, if Request.Form("selOrderItem") is empty, we know that order.asp was opened by the Submit method of the form on specials.html and the AddNew (top) branch is entered.

When the Else branch is entered, the full function "Cadillac" cursor recordset object that you saw in the AddNew part of this chapter is used. We need a full function recordset in order to employ the Delete method. This time, the value set for the *Source* parameter of the Recordset object is the SQL string that you saw in the section of this chapter on dynamic HTML rendering rather than a table name:

```
thisSession = Session.SessionID
    sqlString = "Select * From shopcart WHERE SessionID =" & thisSession
```

Using the SQL string shown in the preceding code as the *Source* parameter value and *adCmdText* as the *Option* parameter opens a very efficient read-write recordset containing order items from the Shopcart table for the current TV viewer. The Delete method of a Recordset object removes the current record from a table on the database. The Move method of a Recordset object is used to set a current record. For example, if you wanted to delete the third record in a recordset, you could use the following code fragment:

```
Recordset.Move 2
Recordset.Delete
```

Notice that the argument given to the Move method, used to move to the third record, is 2. This is because, like most arrays, rows in recordsets are zero-based. The number passed to the Move method in the delete item code is supplied by the value passed to the server by the selOrderItem input of the <form> tag and held on the server in Request.Form("selOrderItem"):

```
Moves = Request.Form("selOrderItem") - 1
objRs.Move Moves
objRs.Delete
```

To understand how Request.Form("selOrderItem") – 1 can provide the correct number of the record to delete, we must think back to how the selOrderItem Select Option Group is dynamically populated. To get the recordset used to populate the selOrderItem select group, the Execute method of the Connection object, objConn, was used with a SQL string to open a recordset on the current TV viewer's records in the Shopcart table in Bob's database. The recordset returned was named objRs. The following structure is used to iterate through objRs and increment the variables named ordertotal, ordercount, and unitcount:

```
<% Do While Not objRs.EOF


    'Loop through records to dynamically build
    'Option group and total units ordered
    'Use ordercount as name value for options
    'So option may be identified as record to remove

    ordertotal = ordertotal + CCur(objRs("PizzaTotal"))
    ordercount = ordercount + 1
    unitcount = unitcount + objRs("PizzaQuan")

%>
```

The variable ordercount is used to fill the value attribute of the <option> input tags in the selOrderItem Select Option Group. Each record in the objRs recordset writes one option into the group using the following code:

```
<option value="<%= ordercount %>">
<%= objRs("PizzaSize")%> Inch <%= objRs("PizzaType") %>
-- <%= objRs("PizzaQuan")%> Pies For <%= objRs("PizzaTotal")%>
</option>
```

The first option, or to put it another way, the first pizza, gets a value of 1 because:

```
ordercount = ordercount + 1
<option value="<%= ordercount %>">
```

That sounds a good deal more complicated than it is. To understand how items are deleted from the shopping cart on the page orders.asp, let's cycle the application, keeping a close eye on what parts tick over. If a TV viewer placed three pizza order items into his shopping cart, there would be three records in the objRs recordset used to dynamically write the selOrderItem Select Option Group. When the first option tag is written, the ordercount variable will be equal to one (*ordercount = ordercount +1*), so the value attribute of the top option will equal one. The second and third options will have values of 2 and 3, respectively. Now suppose the TV viewer selects the option "14 Inch kahuna, 2 Pies $32.78" to delete, and then presses the Remove button. The Remove button's click event calls the setItem() procedure:

```
<input type="button" name="btnremove" value="Remove Item"
onclick="setItem()" >
```

To see the setItem() procedure on order.asp in your text editor, scroll up to the <script></script> tag set at the top of the page. The setItem() procedure is shown in the following code:

```
/* setItem function is called by the Remove Item button*/
function setItem()
{
    var thisform = document.forms[0];
    thisform.submit()
    return;
}
```

The setItem() function submits a form containing the item to be removed from the Shopcart table, in this case, option value 1. The form passes the value 1 to the server where it becomes the top or zero record and gets deleted. The following code shows how this is accomplished:

```
Else

    'Move to selected record and Delete
    'Use SQL string with SessionID
    'So only this customer's records are affected

    objRs.Open sqlString, ,adCmdText
    Dim Moves

    objRs.Move Moves
    objRs.Delete

    End If
```

When order.asp is returned to the client, the shopping cart option group is displayed with one fewer item.

In-Line Empty Recordset Validation

If no item in the select group of options is selected and the TV viewer presses the Remove button, an error is generated. To prevent this kind of error, a small procedure is run when the order.asp page onload event fires and selects an option using the DOM. The name of this function is setOpenOption.

```
<body background="images\contentile.jpg" onload="setopenoption()">
```

An item is selected in the Order box by a procedure that is run when the page first opens:

```
<script>
function setOpenOption()
{
    document.forms[0].selOrderItem.options[0].selected = true
}
</script>
```

There is only one form on order.asp, so it may be identified in the DOM by document.forms[0]. The option group we are concerned with is named selOrderItem. Option groups maintain an array named options representing the items they contain. Thus, selOrder.options[0] represents the top option. Combining the whole bit with the selected method of the option group ensures that somebody in the option group will always be selected.

Testing for the Empty Recordset EOF and BOF

If the TV viewer removes all the items from the Order box, it is important to disable the Remove Item and Next buttons, or the TV viewer could cause an error to be generated. The way to test for an empty recordset is to poll the Beginning Of Field (BOF) and End Of Field (EOF) properties, as shown in the following sample code:

```
<%
If objRs.BOF and objRs.EOF Then

    ordertotal = "None"
    unitcount = "None"

%>
    <option> No Pizzas Currently Ordered </option>
<%

End If
%>
</select> </form>
```

To see the empty record set code in your text editor, scroll down just below the Do Until loop structure that was used to populate the selOrderItem Select Option Group. If all the TV viewer items have been removed from the Shopcart table, an empty recordset is detected by testing the Beginning Of Field (BOF) and End Of Field (EOF) properties. If the recordset is empty, then descriptive text is placed in the select group and the variables ordertotal and unitcount are set to the text string "None". Incidentally, this illustrates the flexibility of un-typed scripting variables. The variables ordertotal and unitcount held numbers earlier in this description. Now they hold text strings and the VBScript interpreter could care less. The value of unitcount is tested to decide if a disabled button should be displayed using the ASP and HTML dynamic rendering technique shown earlier in the chapter:

```
<% If unitcount = "None" Then %>
   <input type="button" name="btnRemove" disabled="True" value="Remove
      Item" onclick="setItem()" >
   <input type="button" name="btnNext" disabled="True" value="Next >>"
      onclick="getcustomer()">
<% Else %>
   <input type="button" name="btnRemove" value="Remove Item"
      onclick="setItem()" >
   <input type="button" name="btnNext" value="Next >>"
      onclick="getcustomer()">
<% End If %>
```

The result of an empty recordset is shown in Figure 20.10.

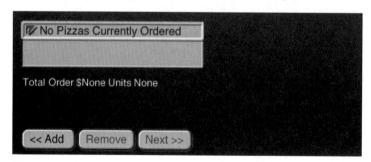

Figure 20.10 *Result of empty recordset.*

Using the Empty Recordset to Validate TV Viewer Login

The empty recordset is commonly used to validate TV viewers on sites that use passwords. For example, suppose Bob had some sort of pizza club and someone joined Bob's pizza club. Member validation could be accomplished by opening Bob's Pizza with a page that asked the TV viewer for two bits of data, such as e-mail address and password. The submit event of this gatekeeper page could open a recordset using a SQL string with a Where clause that filtered the values entered by the TV viewer. If an empty recordset is returned, the TV viewer could be redirected to a sign-up page. If the e-mail and password validated to a record, then another recordset could be opened and used to fill values in the customer information form.

GETTING CUSTOMER DATA WITH THE CUSTOMER.HTML PAGE

When a TV viewer is done adjusting her order, she can press the Next button, fire the getcustomer() procedure (shown shortly), and open the customer.html page. The mission of the customer.html page, shown in Figure 20.11, is to gather information such as name, address, and phone number.

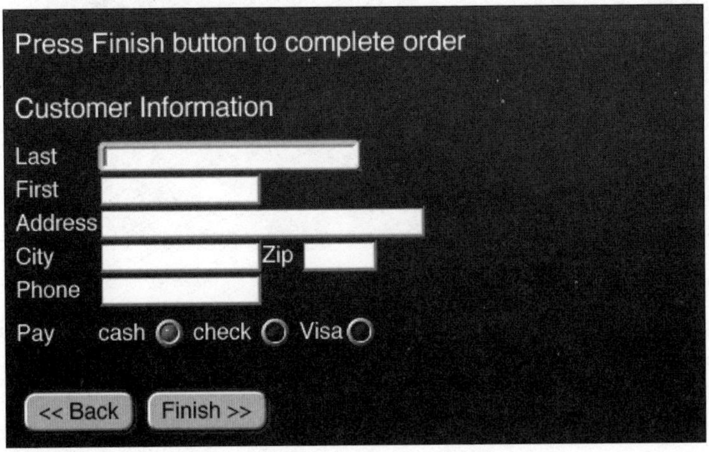

Figure 20.11 *The customer.html page.*

The procedure to open a form in the right frame of Bob's Pizza is as simple as the one shown here:

```
window.location.href = "customer.html"
```

By changing the value of the href property to customer.html, the form page is loaded into the Bob's Pizza content frame. Notice that the extension for the customer data page is .html rather than .asp. No server-side script or dynamic rendering is required to deliver the customer page, so ASP functionality is not needed on this page.

Going Back to Order.asp (The Back Button)

Special care must be taken with a button like Back because the Back button causes an .asp page to be swapped for an HTML page. You have seen many navigation buttons in the course of reading through the examples in this book. Most of these buttons have caused one HTML page to be swapped for another HTML page. In the

Humongous Insurance example, you saw an ASP page being swapped for an HTML page when TV viewers clicked Home on the thanks.asp page. All these HTML to HTML and ASP to HTML navigation methods used the same general syntax in their code:

```
window.location.href = "pagename.html"
```

Unfortunately, you can't use the same code to navigate from an HTML page to an ASP page—at least not if the ASP page is like order.asp. If you use window.location.href = "pagename.html" with the Back button on customer.html to navigate back to order.asp, you see an error like the one shown in Figure 20.12.

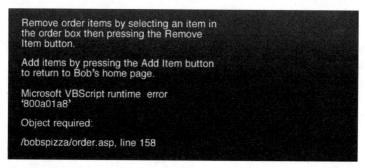

Figure 20.12 *Error shown if wrong method used to open order.asp.*

The reason for the error shown in Figure 20.12 is that order.asp is expecting a form submission from specials.html in order to build its dynamic content. Since the navigation occurs in customer.html, no form field data on specials.html is passed. The hard way of fixing this problem is to pass around the values in the Shopcart table for this TV viewer to all the pages using hidden form fields. The easy way is just to use the following syntax when swapping an ASP file for an HTML file:

```
/* Called by << Back button   */

function goback()
{

    window.history.back(1)

}
```

The History property of a Window object contains a *readonly* reference to the History object. The History object is a *readonly* array of strings that specify the URLs of pages the browser has visited. The Back method of the History object allows backward navigation in the window without causing an error.

CLIENT-SIDE DATA VALIDATION IN MICROSOFT TV

In Chapter 17, "Creating Forms for Microsoft TV Content," you saw how to accomplish server-side validation using VBScript and ASP. It is also possible to validate user input on the client side. Client-side validation is advantageous because it can reduce server load. The next section will show you a simple client-side procedure that validates for empty fields. In your own application, you will want to expand validation to data types, field lengths, and even data accuracy.

> **NOTE** Supported events useful in Microsoft TV client-side validation are *onChange*, *onClick*, and *onBlur*. Unsupported events that are commonly used are *onKeypress*, *onKeyup*, and *onKeydown*.

Validating for Empty Fields

You must guard against the submission of forms with critical fields empty. Unless database fields are set up to accept Null values, an empty field can cause an ODBC error, stopping your page and causing an embarrassing message to be displayed. Even if no ODBC error occurs, missing information can be frictional. After all, if Bob's employees have to call the TV viewer because he left the address field blank, what's the point of an automated system? To guard against missing data, the *onClick* event of the "Finish" button calls the setData function:

```
function setData()
{
    var thisform = document.forms[0]


    if(thisform.txtLastName.value == "" ||
        thisform.txtFirstName.value == "" ||
        thisform.txtAddress.value == "" ||
        thisform.txtCity.value == "" ||
        thisform.txtZipCode.value == "" ||
        thisform.txtPhone.value == "")
    {

        window.self.location.href = "validate.html"
    }
    else
    {
        thisform.submit()

    }

}
```

As you can see, the setData function tests for content in each field on the customer information form. If any of the fields are empty, the TV viewer is prompted to fill in the form by a screen like the one shown in Figure 20.13.

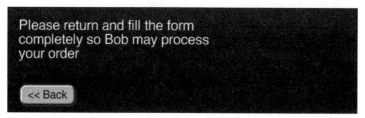

Figure 20.13 *The validate.html page.*

If all the fields have data in them the form can then be submitted, as shown in the following sample code:

```
else
{
    thisform.submit()
}
```

This form submits data that builds the final page in Bob's Pizza, the thanks.asp page. Before going on to explain the details of thanks.asp, let's take a moment to see why some common validation methods are not applicable in Microsoft TV 1.0 coding.

No Alert or Message Boxes in TV Mode

One common way to handle improper TV viewer input in Web development is to put up an alert box like the one shown in Figure 20.14.

Figure 20.14 *A common message box.*

Alert or message boxes such as the one shown in Figure 20.14 cannot be shown in Microsoft TV unless the browser is in Web view. However, most, if not all, interactive TV applications run in TV view, so the TV object can display the video signal. Therefore, messages boxes and alert boxes are not an option for most interactive TV content.

Disabling Buttons

Another popular method used in form input validation is to disable the Next button, (or in this case, the Finish button) until all the fields are filled. To employ this technique, the page is opened with the Finish button grayed out, as shown in the following code:

```
<input disabled="true" type="button" name="btnfinish" value="finish >>">
```

Each time data is entered in a form field, a function is run. If all the fields have data, the Finish button is enabled using the DOM.

```
thisform.btnFinish.disabled = false
```

At the time of this writing, there is a problem with using the DOM to enable a control like Finish in Microsoft TV 1.0. It may be fixed by the time you read this, but for now buttons cannot be fully enabled using the DOM in client-side script. As it currently stands, the button looks gray until it gets focus, even if its disabled attribute is set false. And grayed-out buttons will lead the TV viewer to believe something is wrong.

> **IMPORTANT** As you saw on order.asp, disabling buttons provides a good way to prevent bad TV viewer input on an ASP page. This is because ASP script is run on the server side and it is not subject to client-side DOM implementation issues.

When customer data is submitted to the server, as shown in the following sample code, it is time for the final page in Bob's Pizza, thanks.asp, to do the real heavy lifting in this ASP-ADO story:

```
<form name="frmtotorder" method=post action="thanks.asp">
```

CODING ASP FOR MULTIPLE TABLES ON THANKS.ASP

It is a good idea to give TV viewers some feedback when they are finished ordering a product, as shown in Figure 20.15.

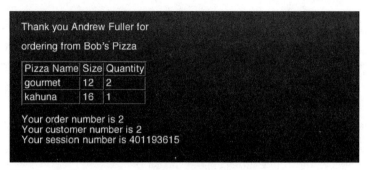

Figure 20.15 *Feedback on the thanks.asp page.*

This section shows how the customer and pizza data accumulated so far is combined and written to the relational database that holds Bob's Pizza order information. Those of you who are old hands with database concepts will have little difficulty understanding what is going on in the ASP-ADO code that makes up thanks.asp. However, for those of you who want a walkthrough of Bob's database and a nickel tour of relational database concepts in general, see the "Bob's Pizza Database Explained" section later in this chapter.

Rendering and Recording Order Information

The information on thanks.asp is rendered in a familiar way. Standard HTML tags are mixed with ADO-driven data written in ASP tags, as shown in the following code sample:

```
Thank you  <%= Request.Form("txtFirstName")%>
<%Response.Write(" ")%> <% = Request.Form("txtLastName")%>
for <br><br> ordering from Bob's Pizza
```

The values for customer name are submitted to the server by the form on the customer data page, customer.html. The ASP page, thanks.asp, has the job of combining all the data collected about a single TV viewer's order and writing it to the appropriate tables in Bob's database. Before seeing how this is accomplished, we need to detour for a moment to discuss the tables in Bob's database, as shown in the following table:

Table	Fields
Customer	CustomerID, Name, Address, Phone, etc.
OrderDetails	OrderID, ProductID, Quantity ordered
Orders	OrderID, Payment method (PayBy), Order date
Products	ProductID, ProductName, UnitPrice, description ID
ProdDesc	ProdDescID, Text description of products
Shopcart	SessionID, PizzaType, PizzaSize, PizzaQuan, PizzaTotal, ProductID

Writing to the Customer Table

To see how data writing is accomplished in code, open thanks.asp in your trusty text editor and scroll down to code that resembles the following:

```
<%

If Not IsEmpty(Request.Form) Then
Dim thisorder
Dim sqlString
Dim thissession
thissession = Session.SessionID
sqlString = "Select SessionID, ProductID, PizzaType, PizzaSize, →
Sum(PizzaQuan) As SumOfPizzaQuan From shopcart Group By SessionID, →
ProductID, PizzaType, PizzaSize Having SessionID=" & thissession

'Open connection to DSN named Bobs on server
Set objConn = Server.CreateObject("ADODB.Connection")
objConn.Open "Bobs"

'Add new customer data to Customer table.

Set objCustomerRs = Server.CreateObject("ADODB.Recordset")
objCustomerRs.ActiveConnection = objConn
objCustomerRs.CursorType = adOpenKeyset
objCustomerRs.LockType = adLockOptimistic
objCustomerRs.Source = "Customer"
objCustomerRs.Open
```

(continued)

```
'Invoke the AddNew method and give values to all the fields
'in the Customer table as represented by the objCustomer
'recordset object

objCustomerRs.AddNew

objCustomerRs("LastName") = Request.Form("txtLastName")
objCustomerRs("FirstName") = Request.Form("txtFirstName")
objCustomerRs("Address") = Request.Form("txtAddress")
objCustomerRs("City") = Request.Form("txtCity")
objCustomerRs("ZipCode") = Request.Form("txtZipCode")
objCustomerRs("Phone") = Request.Form("txtPhone")

'The Update method adds a new row to the Customers table and
'makes it the current record

objCustomerRs.Update

'Fill the variable thiscustomer with the Autonumber created by
'Access for the new record CustomerID field

thiscustomer = objCustomerRs("CustomerID")

'Close and remove the object from memory to free up server resources

objCustomerRs.Close
Set objCustomerRs = Nothing
```

Code that writes to the Customer table should look familiar. The only real difference between it and what we have seen before is that the table being opened is the Customer rather than the Shopcart table.

One field of data in the Customer table is vital to relating order information to the other tables. It is the CustomerID field. The CustomerID number for each new record is generated by a special Access data type called AutoNumber. Notice the last row in the table represented in Figure 20.16.

CustomerID	LastName	FirstName	Address	City	ZipCode	Phone
1	Davolios	Nancy	507 20th E. Apt. 2A	Seattle	98122	206-555-5678
2	Fuller	Andrew	908 W. Capital Way	Tacoma	98401	206-555-8754
(AutoNumber)						

Record: 1 of 2

Figure 20.16 *Bob's Customer table uses AutoNumber for CustomerID.*

The Update method of the Recordset object creates a new row in the Customer table and makes that record the current record. A variable named thiscustomer is filled with the new record's CustomerID field value:

```
objCustomerRs.Update

'Fill the variable thiscustomer with the Autonumber created by
'Access for the new record CustomerID field

thiscustomer = objCustomerRs("CustomerID")
```

The thiscustomer variable is used to write data to the Orders table and relate it to the Customers table.

Writing to the Orders Table

After the data submitted by the customer.asp form is written to the Customer table, the objCustomerRs Recordset is closed and removed from memory. This frees up server resources. Next, the Orders table receives data. The following shows the code used to add new data to the Orders table:

```
'Add new order to orders table

Set objOrderRs = Server.CreateObject("ADODB.Recordset")
objOrderRs.ActiveConnection = objConn
objOrderRs.CursorType = adOpenKeyset
objOrderRs.LockType = adLockOptimistic
objOrderRs.Source = "Orders"
objOrderRs.Open


'Add a new record to the Orders table using the CustomerID value
'held in the thiscustomer variable so the relational join between
'the Customer and Orders tables can be maintained

objOrderRs.AddNew

objOrderRs("CustomerID") = thiscustomer
objOrderRs("Payby") = Request.Form("radPayBy")
objOrderRs("OrderDate") = Now()
objOrderRs.Update
```

(continued)

```
'Fill the variable thisorder with the Autonumber created by
'Access for this new order record

thisorder = objOrderRs("OrderID")

'Close and remove the object from memory to free up server resources

objOrderRs.Close
Set objOrderRs = Nothing
```

The Connection object objConn is recycled to provide the *ActiveConnection* parameter of the objOrderRs Recordset object that will contain the Orders table. The next order number is obtained in the same manner used to find the next customer number, and then the variables thisorder and thiscustomer are used to write data to the OrderID and CustomerID fields. Order date is supplied by the Visual Basic Now() function, which returns the current date and time as supplied by the system clock on the server. Figure 20.17 shows the fields in the Orders table.

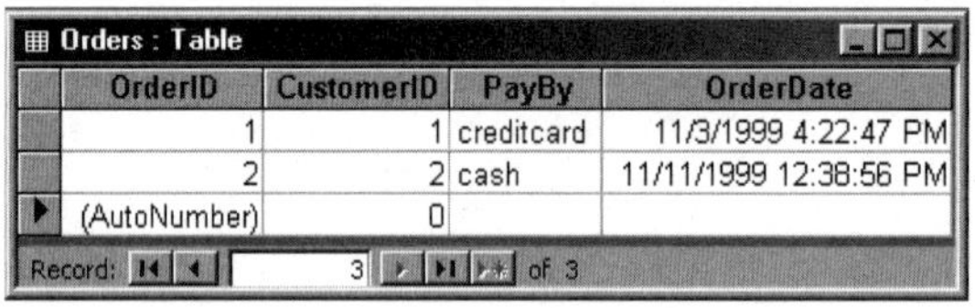
Orders : Table

OrderID	CustomerID	PayBy	OrderDate
1	1	creditcard	11/3/1999 4:22:47 PM
2	2	cash	11/11/1999 12:38:56 PM
(AutoNumber)	0		

Record: 3 of 3

Figure 20.17 *Orders table fields fed by form and function.*

Writing to the OrderDetails Table

The way data is written to the OrderDetails table is a bit less intuitive than the code you have seen so far. First, orders for duplicate product items are combined using a SQL string, then the UpdateBatch method is used to write more than one row to the OrderDetails table in a single operation. The following code shows how this is done:

```
'Set properties of recordset object and open it on table named shopcart
'Use SQL string to open only this customers record and combine pizza types
'There may be many pizzas per order so use UpdateBatch method

'Add Details of Order to the OrderDetails table

Set objOrderDetailsRs = Server.CreateObject("ADODB.Recordset")
objOrderDetailsRs.ActiveConnection = objConn
objOrderDetailsRs.CursorType = adOpenKeyset
objOrderDetailsRs.LockType = adLockOptimistic
objOrderDetailsRs.Source = "OrderDetails"
objOrderDetailsRs.Open
```

```
Set objShopcartRs = Server.CreateObject("ADODB.Recordset")
objShopcartRs.ActiveConnection = objConn
objShopcartRs.CursorType = adOpenKeyset
objShopcartRs.LockType = adLockBatchOptimistic

objShopcartRs.Open sqlString, , , ,adCmdText

Do While Not objShopcartRs.EOF
   objOrderDetailsRs.AddNew
   objOrderDetailsRs("OrderID") = thisorder
   objOrderDetailsRs("ProductID") = objShopcartRs("ProductID")
   objOrderDetailsRs("Quantity") = objShopcartRs("SumOfPizzaQuan")

   objShopcartRs.MoveNext
Loop

objOrderDetailsRs.Updatebatch
objOrderDetailsRs.Close
Set objOrderDetailsRs = Nothing
```

The objOrderDetails Recordset object is opened on the OrderDetails table using the same techniques you have now seen many times. Next, a recordset is opened
on the Shopcart table that combines rows with duplicate ProductID numbers. This
minimizes the number of rows that must be written to the OrderDetails table to
complete one TV viewer's order. Finally, the UpdateBatch method is used on the
objOrderDetails Recordset object to efficiently update the OrderDetails table.

Combining Product Items in SQL

Before we get into the how and why of combining rows from the Shopcart table, let's
take a quick look at Bob's Products table (Figure 20.18) so you can get a handle on
what ProductID numbers mean in terms of pizza.

ProductID	ProductName	UnitPrice	ProductDescID
1	Meat Eater 12"	$12.59	1
2	Meat Eater 14"	$14.50	1
3	Meat Eater 16"	$16.25	1
4	Bob's Gourmet Supreme 12"	$15.25	2
5	Bob's Gourmet Supreme 14"	$17.99	2
6	Bob's Gourmet Supreme 16"	$19.27	2
7	Garden Veggie Delight 12"	$12.25	3

Figure 20.18 *Products table—one ID for each product.*

Product types are represented as ProductID numbers. For example, in Figure 20.18, you can see that ProductID number 4 represents a 12-inch gourmet pizza. We do not want to write a row in the OrderDetails table for multiple ProductID 4 pizzas ordered by a single customer. Instead, we want to write a row for each ProductID the customer wants and increment the quantity field for multiple orders of the same type.

Most of the time you can rely on TV viewers to use the quantity drop-down list box on the pizza specials.html order form to select more than one of a particular pizza. But what if they make a mistake? For example, suppose a viewer had not seen the quantity option group on Bob's order page. He might have sent in his order with the items shown in Figure 20.19.

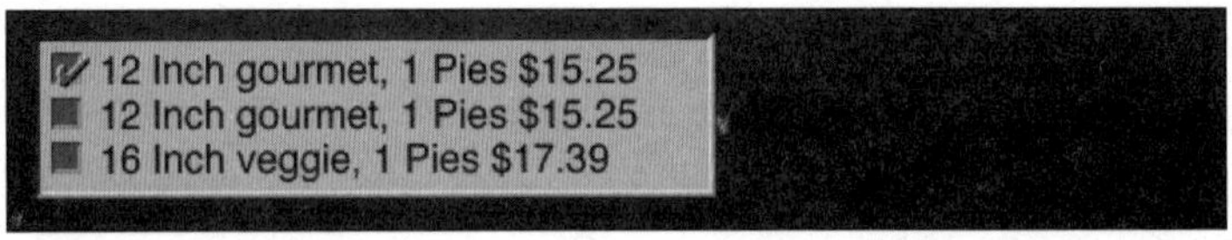

☑ 12 Inch gourmet, 1 Pies $15.25
■ 12 Inch gourmet, 1 Pies $15.25
■ 16 Inch veggie, 1 Pies $17.39

Figure 20.19 *Order with multiple orders for the same product item.*

Recognizing that there are two 12-inch gourmet pizzas, and that the ProductID number is 4 in the order, allows us to combine them into one row in the OrderDetails table, as shown in Figure 20.20.

OrderID	ProductID	Quantity
1	9	2
1	10	1
2	4	2
2	12	1
0	0	0

Figure 20.20 *OrderDetails table.*

Imagine the customer is assigned order number 2 as seen in Figure 20.20. You can see that order number 2 consists of two rows: one for two number 4 pizzas and another row for one number 12 pizza. If the customer's order was written to the OrderDetails table as it was submitted, three rows in the OrderDetails table would have been required for order number 2. That simply would not do.

Therefore, it is essential to combine orders by ProductID in order to minimize the number of rows written. To do this, we use a special sort of SQL string as the *Source* parameter to the Open method when we create the objShopcartRs recordset, as shown in the following sample code:

```
thissession = Session.SessionID
sqlString = "Select SessionID, ProductID, PizzaType, PizzaSize, →
    Sum(PizzaQuan) As SumOfPizzaQuan From shopcart Group By SessionID, →
    ProductID, PizzaType, PizzaSize Having SessionID=" & thissession
```

Access is used to create the following SQL string. As shown in the following code, the objShopcartRs object uses the SQL string held in the variable sqlString as the *Source* parameter in its Open method:

```
objShopcartRs.Open sqlString, , , ,adCmdText
```

As with all the SQL strings used in Bob's Pizza, this one was kept simple enough that it could be built in Access using the QBE grid, as shown in Figure 20.21.

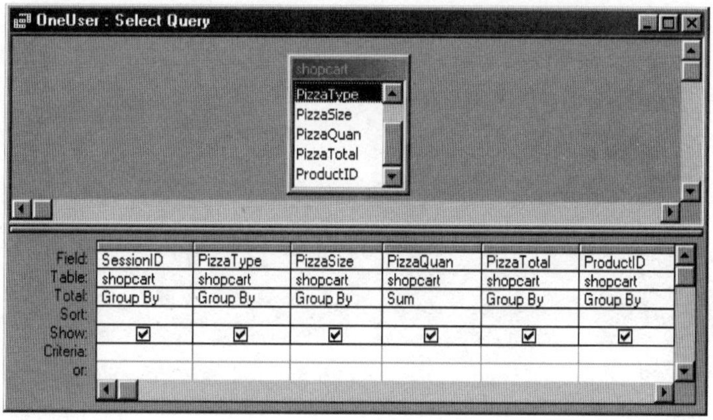

Figure 20.21 *Access QBE grid used to Sum fields in Shopcart table.*

NOTE For more complete information on building simple SQL strings using Access, see the topic "Using Access To Do Your SQL Talking for You" in the "Tips and Tricks in ASP-ADO Coding" section of this chapter.

Notice that the Totals (S) button is depressed in the Access toolbar. This creates a new row (Total) in the QBE grid allowing you to combine records. This step is necessary to handle the situation where a TV viewer orders two of the same ProductID on separate rows. The SQL string created by the QBE grid is accessed by clicking SQL View in the QBE grid View menu.

Using the UpdateBatch Method on the OrderDetails Table

Earlier in this chapter, we discussed how an order ID number such as 2 can contain multiple rows in the OrderDetails table. This is because a customer may order many pizzas in one order. Rather than writing these rows one at a time, it is more efficient to use the UpdateBatch method and write them all at once. The UpdateBatch method allows us to send both rows of a customer's order from the shopcart recordset to the OrderDetails table in one batch. After looping through the shopcart recordset, *objShopcartRs* assembles the records to be sent, and then the UpdateBatch method is called, as shown in the following code.

```
Do While Not objShopcartRs.EOF
    objOrderDetailsRs.AddNew
    objOrderDetailsRs("OrderID") = thisorder
    objOrderDetailsRs("ProductID") = objShopcartRs("ProductID")
    objOrderDetailsRs("Quantity") = objShopcartRs("SumOfPizzaQuan")

    objShopcartRs.MoveNext
Loop

objOrderDetailsRs.Updatebatch
```

Rendering a Table Dynamically

The remainder of the code on thanks.asp loops through the shopcart recordset and displays the result in a table on a screen like the one seen in Figure 20.22.

Figure 20.22 *A table is built on the fly with one row per order item.*

The objShopcartRs and objConn objects are then closed and removed from memory, as shown in the following code:

```
<table border=1 class="description">
<tr><td>pizza name </td><td>size </td><td>quantity</td></tr>
<% Do While Not objShopcartRs.eof %>

    <tr><td><%= objShopcartRs("pizzatype")%></td>
    <td><%= objShopcartRs("pizzasize")%></td>
    <td><%= objShopcartRs("sumofpizzaquan")%></td></tr>


    <%
    objShopcartRs.MoveNext
Loop
objShopcartRs.Close
Set objShopcartRs = Nothing
objConn.Close
Set objConn = Nothing
%>
```

```
</table>
<br><br>
Your order number is <%= thisorder%> Your customer number is <%=
thiscustomer%> Your session number is <%= thissession%><br>
```

Each time through the objShopcartRs recordset, the following code writes a row to the HTML table dynamically:

```
<tr><td><%= objShopcartRs("pizzatype")%></td>
<td><%= objShopcartRs("pizzasize")%></td>
<td><%= objShopcartRs("sumofpizzaquan")%></td></tr>
```

There is another benefit of combining rows in the SQL string used to open the objShopcartRs recordset. Since orders are combined by type, the table will never have more than four rows. Orders are combined by their product type: gourmet, meat eater, veggie, and kahuna.

DATABASE MAINTENANCE

In a production application, regular database maintenance would include clearing out the Shopcart table and running a query to remove duplicate records from the customers table. It is important to understand that some of the data access syntax used in this example would not scale to the millions of simultaneous hits that a national ad campaign might generate. If you are planning an ad campaign that will generate extreme amounts of traffic, you need to consult database professionals. They will use less cursor functionality and more complex SQL string language than you can create in the Access QBE grid.

BOB'S PIZZA DATABASE EXPLAINED

The global theory and practice of relational database design is beyond the scope of this book. There are literally thousands of pages written on this subject alone. That said, here is just enough background in relational database theory to help you understand what is going on in Bob's database.

Relational Database Tables

The big job of thanks.asp is to write order information to the relational tables on Bob's database. The relationship between the tables in Bob's database is shown in Figure 20.23.

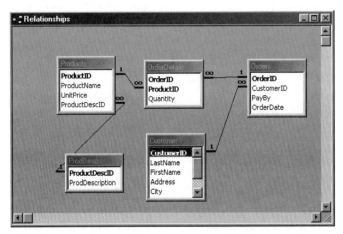

Figure 20.23 *Relationship between tables in Bob's database.*

What's Relational About a Relational Database?

The short answer to "What is relational about a relational database?" is: duplication of information is kept to a minimum by relating tables to one another through some shared field. For example, in Bob's database, there are a Customer table and an Orders table. Each customer could have many orders. Rather than writing a customer's information, such as name, address, and phone number, each time he makes a new order, the Orders table simply notes a CustomerID number. The CustomerID number in the Customer table has a special status. It is called the primary key. In the Orders table, the CustomerID is just another field. When the tables are related it is called a "join." More specifically, when the primary key of a table is related to a "foreign" key of another table, the join is a "one-to-many" join.

One-to-many joins are represented by a 1 and an infinity symbol in the Relationships window, as shown previously in Figure 20.23. This is all well and good because one customer can have many orders in the database. When the TV viewer wants to see information about the orders they have entered, the Customer and Orders tables are melded into a combination table called a query. Queries are used as the basis for Forms and Reports, which combine information in a useful and graphically pleasing way. For example, Access forms can combine customer and order information on one screen and Reports may be used as printed invoices.

Using Northwind as a Template

The table structure used in Bob's did not just jump out like an epiphany on the road to Redmond. It was copied wholesale from the Northwind sample database and changed to fit the needs of Bob's Pizza. The Northwind sample database was created by those brainy guys at Microsoft in the Access group, who all wear size 14 hats and eat fish for practically every meal. It has been the premier Access sample database for years, so you can bet that the table structure in Northwind is "spot on." When you go to create your own underlying database, try to morph the Northwind sample database to the needs of your customer or department. That is what it is there for. If you've ever bought Microsoft Access, Microsoft Office, Visual Basic, or a score of other products, you have a copy of the Northwind database. To find it, search on *.mdb. You may find the database file to be named Nwind or Northwind, depending on its host application.

Bob's Table Structure

There are three tables that need to receive data in order to complete an order for Bob's Pizza. They are the Customer, Orders, and OrderDetails tables. The Customer table receives name, address, and phone information. The Orders table holds CustomerID, OrderID, order date, and information on how the order will be paid for. Finally, the OrderDetails table holds which pizzas are ordered and how many of each. So each order will contain one row in the Customer and Orders tables and any number of rows in the OrderDetails table. The great thing is that only the most minimal information is repeated in the three tables.

For example, imagine that an employee of the now famous Northwind Trading Company, named Andrew Fuller, decided to order a couple of pizzas. The Customer and Orders tables would each receive a row of data, as shown in Figures 20.24 and 20.25.

	CustomerID	LastName	FirstName	Address	City	ZipCode	Phone
▶	1	Davolios	Nancy	507 20th E. Apt. 2A	Seattle	98122	206-555-5678
	2	Fuller	Andrew	908 W. Capital Way	Tacoma	98401	206-555-8754
＊	(AutoNumber)						

Figure 20.24 *Customer table from Bob's Pizza.*

	OrderID	CustomerID	PayBy	OrderDate
▶	1	1	creditcard	11/3/1999 4:22:47 PM
	2	2	cash	11/11/1999 12:38:56 PM
＊	(AutoNumber)	0		

Figure 20.25 *Orders table from Bob's Pizza.*

The main thing to notice is that the CustomerID field in the Customer table matches the CustomerID field in the Orders table (2). The number and type of pizzas ordered is written to the OrderDetails table, as shown in Figure 20.26.

OrderDetails : Table

OrderID	ProductID	Quantity
1	10	1
1	12	1
2	4	2
2	9	1
0	0	0

Record: 4 of 4

Figure 20.26 *OrderDetails table from Bob's Pizza.*

Notice that the OrderID 2 appears once in the Orders table and twice in the OrderDetails table. Looking at the Quantity and ProductID fields, you can see that the customer wants two number 4 and one number 9 pizzas. So what's a number 9 pizza? Let's take a look at the relational Products table in Figure 20.27 for an answer.

Products : Table

ProductID	ProductName	UnitPrice	ProductDescID
1	Meat Eater 12"	$12.59	1
2	Meat Eater 14"	$14.50	1
3	MeatEater 16"	$16.25	1
4	Bob's Gourmet Supreme 12"	$15.25	2
5	Bob's Gourmet Supreme 14"	$17.99	2
6	Bob's Gourmet Supreme 16"	$19.27	2
7	Garden Veggie Delight 12"	$12.25	3
8	Garden Veggie Delight 14"	$14.56	3
9	Garden Veggie Delight 16"	$17.56	3
10	Big Kahuna Hawaiian 12"	$13.33	4
11	Big Kahuna Hawaiian 14"	$16.75	4
12	Big Kahuna Hawaiian 16"	$19.95	4
(AutoNumber)		$0.00	0

Figure 20.27 *Products table from Bobs.mdb.*

As shown in the preceding illustration, you can see from the Products table that a number 4 is a Bob's Gourmet Supreme 12" and a number 9 is a Garden Veggie Delight 16". Notice that pizzas are organized by type and size at Bob's. This means that descriptions of the products would repeat if they were kept in the Products table. After all, a 12-inch Gourmet Supreme has the same set of ingredients as a 16-inch Gourmet Supreme. Since repeating data in two or three rows would amount to relational database blasphemy, pizza descriptions are kept in their own table joined to the Products table by the ProductDescID, as shown in Figure 20.28.

ProdDesc : Table

ProductDescID	ProdDescription
1	Peperoni, bacon, ham, mushrooms, onions, on a double thick dough boy crust
2	Sausage, olives, mushrooms, peppers, tomatoes, garlic & pesto on a garlic crust
3	Organic goat cheese, sun dried tomatoes, sliced garlic and pine nuts on a whole wheat crust
4	Canadian bacon, pineapple, mushrooms, sprinkled with a crispy macadamia nut crust
(AutoNumber)	

Record: 1 of 4

Figure 20.28 *Product Description table from Bob's Pizza.*

As you have seen, the table structure in Bob's database is designed to record orders with a minimum amount of data. Wherever possible, the data is written as integer values. These design items keep server load to a minimum and database performance maximized.

The only trouble with writing most of the order data as integer values is that the tables then contain little human-readable information. In a fully fleshed-out example, the tables in Bob's database would be combined into queries. The queries would be the basis for human-readable Access forms and reports. It is beyond the scope of this book to teach you how to create Access forms, but you can get plenty of good examples by looking at how it is done in the Northwind database. Likewise, you can read a book about Access.

TIPS AND TRICKS IN ASP-ADO CODING

This section will arm you with some of the tips and tricks, as well as references, that e-commerce Web developers use when working with ADO, ASP, and SQL.

The Object Browser

The best single source for information on Microsoft object models is the Object Browser. In the Object Browser, you can find help topics on the members, methods, properties, and constants available in object families such as ActiveX Data Objects (ADO).

As an example of how to use the Object Browser, the next section explains how you might replace constants like *adOpenTable* with their numeric equivalents. Numeric equivalents for constant values can make your code more bulletproof against things like the SSI file adovbs.inc getting lost or corrupted.

Replacing ADO Constants with Their Numeric Equivalents

You can view the numeric values for constants like *adOpenKeyset* by using the Object Browser. Simply open the Visual Basic Editor (VBE) from Microsoft Word, Microsoft Excel, Access, or Microsoft Visual Basic.

To view topics on a particular Object model such as ADO, you must establish a reference to its DLL. To do this you check the references dialog box in the Visual Basic Editor (VBE). The VBE is part of many Microsoft development products, such as Office, Visual Interdev, and Visual Basic. For example, to open the VBE in Word, point to Macro on the Tools menu and click Visual Basic Editor.

In the VBE, click References under the Tools menu of Word, Excel, or Microsoft PowerPoint. Make sure the Microsoft ActiveX Data Objects (ADO) Library has been selected in the Available References list. Close the Reference dialog box.

To open the Object Browser, in the editor window, click Object Browser in the View menu. To see topics on ADO objects, search on ADO in the Object Browser, as shown in Figure 20.29

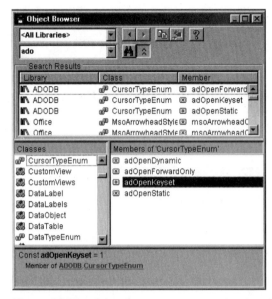

Figure 20.29 *Object Browser numeric value for* adOpenKeyset *constant.*

In the previous illustration, notice in the lower left corner of the Object Browser window that the numeric value of 1 is equal to the *adOpenKeyset* constant. In code you could use the value "1" in place of the *adOpenKeyset* constant to bulletproof your code against corruption or loss of the adovbs.inc file.

Using OLEDB Instead of ODBC DSN

Depending on what version of Jet (Access) and IIS you are running, it may be possible to connect to your data source using native OLEDB technology without using the OLEDB driver to the ODBC layer as shown in this chapter's examples. For example, if you are running ADO 2.0 and Jet 4.0, you may be able to increase efficiency by connecting with code that resembles the following:

```
<%
Set objConn = Server.CreateObject("ADODB.Connection")
objConn.Open "Provider=Microsoft.Jet.OLEDB.4.0;" _
& "Data Source=.\Bobs.mdb;"

%>
```

Check with your database administrator or the following Microsoft Web sites for more information on data connection strings:

- *http://www.microsoft.com/data/default.htm*

- *http://www.microsoft.com/data/oledb/default.htm*

- *http://www.microsoft.com/data/ado/default.htm*

- *http://www.microsoft.com/data/odbc/default.htm*

Using Access to Do Your SQL Talking for You

Several times in this chapter, we demonstrated how you can use SQL strings to perform database functions. One of the great database tips of all time is that Microsoft Access can be used to create SQL strings for you, so you can avoid learning cryptic SQL stuff like what an Inner Join is! For example, earlier in this chapter, we discussed how a SQL string was used to return a recordset containing only one TV viewer's shopping cart choices, as shown in the following sample code:

```
thisSession = Session.SessionID
sqlString = "Select * From shopcart where SessionID =" & thisSession

Set objRs = objConn.Execute(sqlString,,adCmdText)
```

To create an equivalent to the SQL string held in the variable sqlString:

1. Open Bob's Access database, named Bobs.mdb.

2. Open a new QBE grid by clicking New on the Queries tab of the database window the OK with the default, Design View type selected in the New Query dialog box.

3. Add the Shopcart table by selecting shopcart from the Show Table dialog box, clicking Add, and then closing the dialog by clicking Close.

4. Add all the fields from the Shopcart table to the QBE grid by selecting and dragging each one.

5. Run the query by choosing Run from the Query menu and copying a value from the SessionID field.

6. Switch back to design view by selecting Design View from the View menu.

7. Set the query up to return records from just one TV viewer by pasting the SessionID number that you copied in step 5 into the Criteria row of the SessionID column, as shown in Figure 20.30.

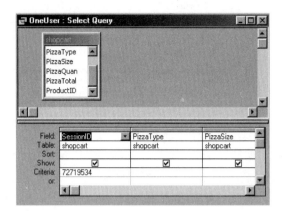

Figure 20.30 *Access QBE grid in design view.*

8. If you like, run the query using directions given in step 5. Notice that only the records with a single TV viewer's SessionID value 72719534 are returned.

9. View the SQL string that Access created by selecting SQL View from the View menu, as shown in Figure 20.31.

Figure 20.31 *SQL string created \by Access.*

The SQL string shown in Figure 20.31 may be cut to the Windows clipboard and pasted right into your code. In many cases, the SQL strings you get from Access may be successfully used as is. In the case at hand, the SQL string will need some modification to work with Bob's Pizza. First, the hard-coded value used with the Where clause in the SQL statement must be replaced with the value returned by the ASP Session objects SessionID property, as shown in the following code:

```
"select shopcart.SessionID, shopcart.PizzaType, shopcart.PizzaSize, " _
& "shopcart.PizzaQuan, shopcart.PizzaTotal, shopcart.ProductID " _
& "from shopcart where shopcart.SessionID=" & Session.SessionID
```

To concatenate variable values to a SQL string, enclose it in quotation marks and use the & operator. The SQL string shown in the previous code is equivalent to the one used with Bob's Pizza. However, like all code generators, Access threw in some syntax that will work but is also ugly. Notice that every field name is preceded by the table name, Shopcart. Since only one table is involved, we may use the SQL From clause to clean things up a bit. The following SQL string is equivalent to the one shown previously:

```
"Select SessionID, PizzaType, PizzaSize, PizzaQuan, PizzaTotal, " & _
"ProductID From shopcart Where SessionID =" & Session.SessionID
```

> **NOTE** SQL clause names such as From and Where are not case sensitive, but data fields are. For example both "WHERE" and "where" will work, but "PizzaType" works, while "PIZZATYPE" fails.

OK, so maybe Access doesn't do all your SQL talking for you, but it sure helps! Finally, since we are using all the fields in the Shopcart table, our SQL string may be refined by using the all (*) symbol in our code. The following SQL string is equivalent to both its predecessors and the one used in Bob's Pizza:

```
sqlString = "Select * From shopcart WHERE SessionID =" & Session.SessionID
```

When Should You Upgrade from Access to SQL Server?

Many TV viewers will be able to simultaneously work with a database. The only real limits are system resources and the ability of the Jet database engine to handle contiguous TV viewers. Hard numbers on the ability of the Jet engine to handle multiple TV viewers are difficult to track down but, if more than 60 TV viewers simultaneously try to update the Shopcart table at the same nanosecond, trouble could occur. Since it would take thousands, maybe tens of thousands, of TV viewers to cause 60 or more writes to the database at any given moment, Bob's Pizza will run out of delivery people and oven space before having to upsize the database management system. If you have an application designed to attract customer orders like "mayflies to a search beacon," you may wish to go all the way and build a Microsoft SQL Server back end.

Server-Side Browser Detection and the Browscap.ini file

In these early, heady days of interactive television, developers are faced with a dizzying and expanding array of new and specialized browsers. Even with ATVEF in place, many of the browsers built into various manufacturers' products will be quirky devils, supporting some things in DHTML but not others.

The more interactive TV development you do, the more eccentric and proprietary browsers you are likely to encounter. One of the problems faced by all sorts of Web developers is that not all browsers are created equal. Some tend to ignore tags that other browsers support. This behavior is bad enough in the bifurcated world of Netscape vs. Internet Explorer. In these early days of interactive TV adoption, developers are faced with an even more dizzying array of proprietary browsers. Until things settle down, every brand of set-top box will contain a browser of uncertain parentage, often with eccentric functionality. Even when these browsers are ATVEF compliant, there is room for wiggle; it will be useful to track the peculiar behavior of various browsers. This can be accomplished with alacrity using the browscap.ini file.

Browser information is sent automatically with the HTTP header, and ASP provides a way to detect the browser on the server side. ASP also provides a way to keep .ini files on the browsers you expect to deal with. In many cases, browser manufacturers supply browscap.ini files on their products. This gives the developer a way to get a really granular way of handling browser capabilities.

To use the browscap.ini file, you first create an MSWC.BrowserType object, and then use its browser method to poll the browscap.ini file for detailed information on various browsers. You can then use response.write to render appropriate code line by line, as shown in the following code sample:

```
<!-- If the browser is not IE based the Fieldset tag is meaningless.
   The response.write method is used to add an HR tag instead. -->
<%
Set objBrowscap = Server.CreateObject("MSWC.BrowserType")
If objBrowscap.browser <> "ie" Then
    str = "<hr>"
    response.write(str)
End If
%>
```

The Browser capabilities component (MSWC.BrowserType) is one of the special server component objects that ship with ASP. It consists of a Dynamic Link Library (browscap.dll) and a text file (browscap.ini). You do not have to know anything about the DLL except that it must be in the same directory as browscap.ini. Both files are typically installed in the C:\WinNTRoot\System32\Inetsrv directory on IIS 4.0. Users of IIS 3.0 should look in C:\WinNT\System32\inetsrv\asp\cmpnts. The browscap.ini file contains detailed information about known browsers and a default section, as shown in Figure 20.32.

You can even get updated browscap.ini files through the following URL:
http://www.microsoft.com/ISN/faq/latest_browscapini_file.asp

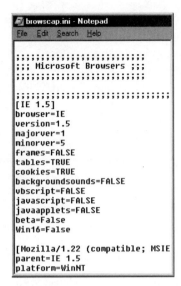

Figure 20.32 *The browscap.ini file.*

The really great thing about browscap.ini files from the interactive TV developer's point of view is that it is easy to catalog the peculiarities of various browsers you will encounter simply by editing the file in a text editor like Windows Notepad. Each bracketed line in the file begins a definition for a particular browser. For example, [IE 1.5] begins the section on Microsoft Internet Explorer 1.5. Each line of the browscap.ini file makes information available to the capabilities components.

For example, let's say you are hoping to use the new <fieldset> and <legend> tags to surround and label a form. Many browsers do not recognize these Internet Explorer–only tags, so putting in an <hr> tag and <fieldset> and <legend> tags will be safe. But suppose you run across a TV receiver based on Internet Explorer. It will render <fieldset>, <legend>, and <hr> tags, making a semi-ugly mess, unless you dynamically tell the browser otherwise, as shown in the following sample code:

```
<Fieldset id="fldCustData">
<legend>Customer Information of Love</legend>
<br>

<!-- If the browser is not IE-based the Fieldset tag is meaningless.
The response.write method is used to add an HR tag instead. -->
<%
Set objBrowscap = Server.CreateObject("MSWC.BrowserType")
If objBrowscap.browser <> "ie" Then
    str = "<hr>"
    response.write(str)
End If
%>
```

The rendering of an <hr> tag is an admittedly trivial use of browser capability technology, but, looking to the .ini file, you should see useful possibilities in polling things like frames and versions. As you develop interactive TV content, you will most likely discover new things that various interactive TV browsers will and will not do. You may catalog them in the .ini file rather than keeping it all rote in your head.

WHAT'S NEXT

This chapter covered a lot of ground, so a quick summary may be in order before moving on. The main ideas to take away from Bob's Pizza are:

- Information is passed from page to page using <form> elements.

- Data for a shopping cart may be assembled in a table.

- Form input information is held by ASP in the Request.Form collection.

- You can create dynamic content by mixing <%%> ASP values with HTML tags.

In the next chapter, "ATVEF and Content Creation Standards," we'll take an inside look at the meaning of the ATVEF specification and then show you where you can find the detailed reference information you need to create robust interactive TV content.

Part V

Microsoft TV Programmer's Guide

Chapter 21

ATVEF and Content Creation Standards

In This Chapter

■ Why You Need the "Microsoft TV Programmer's Guide"

■ References

The Advanced Television Enhancement Forum (ATVEF) is a cross-industry group formed to specify a single public standard for delivering HTML-based interactive TV content that can be authored once and deployed to a variety of television, set-top, and PC-based receivers. At the time of this writing, the ATVEF has developed a Version 1.1 foundation specification that is in the process of approval by its 14 Founder companies. The purpose of the ATVEF Specification, which is available at *http://www.atvef.com*, is threefold:

■ It serves as a guideline for manufacturers of interactive TV receiver devices, including set-top boxes, DVD players, and integrated TVs, and enables manufacturers to build devices capable of receiving interactive TV content created to ATVEF standards.

■ It serves as a guideline for those companies building hardware and software tools for the interactive TV industry, to ensure that the tools they create can be used for authoring and sending content created to ATVEF standards.

■   It serves as a guideline for interactive TV content providers, to ensure that the content they create can be reliably delivered to a wide variety of interactive TV receivers, including set-top boxes, DVD players, and integrated TVs.

WHY YOU NEED THE "MICROSOFT TV PROGRAMMER'S GUIDE"

The ATVEF Specification provides solid guidelines for creating and delivering interactive TV content, but it also leaves a wealth of details to interpretation. For example, the ATVEF Specification 1.1 states that the foundation for ATVEF content is existing Web standards and that mandatory support is required for the following:

■   Hypertext Markup Language (HTML 4.0)

■   Cascading Style Sheets level 1 (CSS1)

■   ECMA 262 Language specification (ECMAScript)

■   Level 0 Document Object Model (DOM 0)

At face value, the preceding bulleted items seem to offer a straightforward guideline for developing Web content for interactive TV. The devil, however, is in the details. For example, at the time of this writing, Microsoft TV supports most HTML elements and attributes, but it does not fully support all HTML 4.0 attributes. In addition, Microsoft TV supports most CSS Level 1 properties, but it does not support all of them.

The ECMAScript and DOM 0 bulleted items in the ATVEF Specification also provide an interesting wrinkle. That's because there is no official Document Object Model 0 recommendation from the W3C, or anywhere else for that matter. In fact, the W3C offers a recommendation only for DOM 1, which led the ATVEF authors to note that DOM 0 and ECMAScript are equivalent to JavaScript 1.1.

In light of the amorphous quality of the ATVEF Specification, combined with the evolving nature of ATVEF support provided by WebTV and other emerging interactive TV vendors, the "Microsoft TV Programmer's Guide," which you can find on the companion CD, has been created for two main purposes:

■   It provides a full reference for HTML 4.0, CSS Level 1, and the Document Object Model 0 (the object model used for JavaScript 1.1) and fleshes out in detail the bulleted items in the ATVEF Specification. The "Microsoft TV Programmer's Guide" also documents the subset of DHTML supported by Microsoft TV.

■ The "Microsoft TV Programmer's Guide" clearly documents what aspects of HTML 4.0, CSS Level 1, CSS Level 2, and the Document Object Model are supported by Microsoft TV. As a baseline for HTML 4.0 and CSS Level 1 support, the "Microsoft TV Programmer's Guide" uses Microsoft Internet Explorer 4.0. As a baseline for the Document Object Model, the "Microsoft TV Programmer's Guide" documents the JavaScript 1.1 object model.

REFERENCES

In addition to the "Microsoft TV Programmer's Guide" chapters in this book and the ITV Programmer's Reference section on the companion CD, you can find more information about interactive TV content standards at the following locations:

■ "Part II: "Microsoft TV Design Guide" in this book

■ The W3C Recommendation for HTML 4.0 at *http://www.w3.org/TR/REC-html40*

■ The W3C Recommendation for Cascading Style Sheets at *http://www.w3.org/pub/WWW/TR/REC-CSS1*

■ The ECMAScript Programming Language Standard at *http://www.ecma.ch/stand/ecma-262.htm*

■ Netscape's JavaScript 1.1 Reference at *http://home.netscape.com/eng/mozilla/3.0/handbook/javascript/index.html*

HTML 4.0 for Microsoft TV

In This Chapter

■   Using the "Microsoft TV Programmer's Guide"

■   Using the "Microsoft TV Design Guide"

■   Using <div> Tags for Positioning

■   About HTML and DHTML

■   References

Specification 1.1 of the Advanced Television Enhancement Forum (ATVEF) defines HTML 4.0 as the standard for building interactive TV content. However, at the time of this writing, Microsoft TV version 1.0 supports perhaps 85 percent of HTML 4.0. While Microsoft TV's HTML support enables you to create just about any type of interactive content, it does have limitations.

For comprehensive documentation about the HTML elements and properties that are supported by Microsoft TV, you can refer to the "Microsoft TV Programmer's Guide" on the companion CD, as described in the following section.

USING THE "MICROSOFT TV PROGRAMMER'S GUIDE"

The "HTML 4.0 for Microsoft TV" topic of the "Microsoft TV Programmer's Guide" on the companion CD is a valuable resource when creating interactive TV content for Microsoft TV. The "HTML 4.0 for Microsoft TV" topic lists the HTML elements, documents their properties, and then highlights the elements and properties that are not supported by Microsoft TV.

USING THE "MICROSOFT TV DESIGN GUIDE"

The "Microsoft TV Programmer's Guide" on the companion CD is designed to be used in conjunction with "Part II: Microsoft TV Design Guide" of this book and with the "Microsoft TV Design Guide" section of the companion CD. The chapters in Part II of this book demonstrate proven strategies and samples for designing interactive TV content using Microsoft TV's HTML, Cascading Style Sheets level 1 (CSS1), and JavaScript support. The "Microsoft TV Design Guide" section of the companion CD contains the source code for the samples. You can use the source code on the companion CD as a starting point for building your own interactive TV content, or you can copy bits and pieces of the sample code as needed.

USING <DIV> TAGS FOR POSITIONING

Perhaps the most important thing to remember when creating interactive TV content for Microsoft TV 1.0 is to position content on the page using <div> tags. For many HTML elements, including the <table> element, Microsoft TV does not support the class or style attributes. As a result, the best strategy for precise positioning of interactive TV content is to enclose an element, or elements, in a division using <div> tags—which support class and style properties—and then position the division by assigning a class to the division or by specifying an inline style, as shown in the following example:

```
<html>
<head>
<title>sample</title>
</head>
<body>
<div style="position:absolute;top:16;left:16">
<table height="300" width="300">
<tr>
<td>sample table</td>
</tr>
</table>
</div>
```

ABOUT HTML AND DHTML

One of the benefits of HTML 4.0 and the associated Microsoft Internet Explorer 4.0 dynamic HTML (DHTML) object model is that each element on a Web page can be referenced as an individual object and dynamically formatted, moved, or hidden. However, Microsoft TV supports only a limited subset of the Internet Explorer 4.0 DHTML object model. Instead, Microsoft TV fully supports standards adopted by the ATVEF, Level 0 Document Object Model (Level 0 DOM) and ECMA 262 Language specification (ECMAScript), which are functionally equivalent to JavaScript 1.1.

REFERENCES

In addition to the "Microsoft TV Programmer's Guide" on the companion CD, you can find more information about interactive TV content standards in the following locations:

- "Part II: Microsoft TV Design Guide" of this book

- The Microsoft WebTV developer site, located at *http://developer.webtv.net*

- The Recommendation site of the World Wide Web Consortium (W3C), located at *http://www.w3.org/TR/REC-html40*

WHAT'S NEXT

This chapter briefly discussed Microsoft TV's support for HTML 4.0. This chapter also stressed the importance of using the "Microsoft TV Programmer's Guide" on the companion CD as a reference source for creating content for Microsoft TV. The next chapter takes a look at Microsoft TV's support for CSS1.

CSS Support for Microsoft TV

In This Chapter

■ Using the "Microsoft TV Programmer's Guide"

■ Using the "Microsoft TV Design Guide"

■ Using CCS Properties to Format Content

■ Microsoft TV CSS1 Quick Reference

■ Microsoft TV CSS2 Quick Reference

■ CSS References

Microsoft TV 1.0 provides support for essential Cascading Style Sheets level 1 (CSS1) properties and a few CSS2 properties, which makes it possible to format interactive TV content using inline styles, embedded styles, or linked style sheets. Nevertheless, Microsoft TV does not provide complete support for all CSS1 properties and attributes. Developers should be aware of which CSS1 properties are, or are not, supported by Microsoft TV. Comprehensive documentation of Microsoft TV's CSS support is presented in the "CSS for Microsoft TV" topic in the "Microsoft TV Programmer's Guide" on the companion CD.

This chapter covers:

■   The "Microsoft TV Programmer's Guide," which provides comprehensive documentation of the CSS properties that are supported by Microsoft TV

■   An important tip about using CSS1 properties only with <div> and <font> tags

■   The Microsoft TV CSS1 Quick Reference

USING THE "MICROSOFT TV PROGRAMMER'S GUIDE"

The "CSS1 for Microsoft TV" section of the "Microsoft TV Programmer's Guide" on the companion CD is a valuable resource for creating interactive TV content. This component of the Programmer's Guide uses Microsoft Internet Explorer 4.0 CSS1 support as a baseline. It documents the properties and attributes that are supported by Microsoft TV, and it indicates those properties and attributes that Microsoft TV does not support.

USING THE "MICROSOFT TV DESIGN GUIDE"

The "Microsoft TV Programmer's Guide" is designed to be used in conjunction with "Part II: Microsoft TV Design Guide" in this book and with the "Microsoft TV Design Guide" section of the companion CD. The chapters in the "Microsoft TV Design Guide" portion of the book offer strategies for using Microsoft TV's HTML, CSS, and JavaScript support to design interactive TV content. The "Microsoft TV Design Guide" on the companion CD contains the source code for the samples covered in the book.

USING CSS PROPERTIES TO FORMAT CONTENT

Use of cascading style sheets to format content gives interactive TV developers more control over the style and layout of a document. However, developers must be aware that there are restrictions when implementing CSS properties for Microsoft TV.

Restrictions on the Use of CSS1 Properties

Follow the important rule shown below whenever you apply CSS properties to interactive TV content for Microsoft TV:

When designing for Microsoft TV, limit CSS properties to <div> and <font> tags.

If you apply CSS properties to HTML elements other than <div> and <font> tags, the properties may not work.

For more information on supported CSS properties, see the table at the end of this chapter or the "CSS for Microsoft TV" topic in the "Microsoft TV Programmer's Guide" on the companion CD.

Formatting Content Using CSS1 Properties: Inline Styles, Embedded Styles, and Linked Style Sheets

With Microsoft TV, you can format interactive TV content using CSS properties in one of three ways:

- Using inline styles
- Using embedded styles
- Using linked style sheets

For more information about implementing inline styles, embedded styles, or using linked style sheets, see Chapter 8, "Formatting Microsoft TV Content with Styles and Style Sheets."

Dynamically Applying Styles

Although the ATVEF does not adopt specifications that require support for dynamic HTML, Microsoft TV 1.0 provides support for a subset of DHTML objects and properties that can be used to perform relatively simple operations such as hiding and showing divisions within a document.

For more information about the subset of DHTML elements and properties supported by Microsoft TV, see Chapter 25, "DHTML for Microsoft TV," and the "DHTML for Microsoft TV" topic in the "Microsoft TV Programmer's Guide" on the companion CD.

MICROSOFT TV CSS1 QUICK REFERENCE

The following table provides a quick reference of the CSS1 properties that are supported by Microsoft TV. For more detailed information about supported CSS1 properties, see the "CSS1 for Microsoft TV" topic in the "Microsoft TV Programmer's Guide" section on the companion CD.

CSS Property	Supported by Microsoft TV?
Font and Text Properties	
font-family	Yes
font-style	Yes
font-variant	No
font-weight	Yes
font-size	Yes
font	Yes
letter-spacing	No
line-height	No
text-decoration	Yes
text-transform	No
text-align	Yes
text-indent	No
vertical-align	Yes
word-spacing	No
Color and Background Properties	
color	Yes
color-background	Yes
background-image	Yes
background-repeat	No
background-attachment	No
background-position	No
background	Yes
Layout Properties	
margin-top	No
margin-right	No
margin-bottom	No
margin-left	Yes
margin	No
padding-top	No
padding-right	No
padding-bottom	No

CSS Property	*Supported by Microsoft TV?*
padding-left	No
padding	No
border-top-width	No
border-right-width	No
border-bottom-width	No
border-left-width	No
border-width	No
border-top-color	No
border-right-color	No
border-bottom-color	No
border-left-color	No
border-color	No
border-top-style	No
border-right-style	No
border-bottom-style	No
border-left-style	No
border-style	No
border-top	No
border-right	No
border-bottom	No
border-left	No
border	No
width	Yes
height	Yes
float	No
clear	No
white-space	No
Classification Properties	
display	No
list-style-type	No
list-style-image	No
list-style-position	No
list-style	No
Units	
length units	Yes
percentage units	Yes
color units	Yes
URL	Yes

MICROSOFT TV CSS2 QUICK REFERENCE

The following table provides a quick reference to the CSS2 properties that are supported by Microsoft TV. For more detailed information about supported CSS2 properties, see the "CSS2 for Microsoft TV" topic in the "Microsoft TV Programmer's Guide" section on the companion CD.

CSS Property	Supported by Microsoft TV?
left	Yes
position	Yes
top	No
visibility	Yes
z-index	Yes

CSS1 REFERENCES

For more information about CSS1, see:

- WebTV Plus Support for Cascading Style Sheets Level 1 at *http://developer.webtv.net/authoring/css*

- The W3C Recommendation for Cascading Style Sheets at *http://www.w3.org/TR/REC-CSS1*

WHAT'S NEXT

This chapter discussed Microsoft TV's support of CSS. It also provided a series of reference points where you can get more information about CSS. The next chapter covers the Document Object Model (DOM) used for Microsoft TV and demonstrates client-side scripting using JavaScript 1.1.

Chapter 24

Document Object Model for Microsoft TV

In This Chapter

■ The "Microsoft TV Programmer's Guide"

■ Using JavaScript with Microsoft TV

■ Unimplemented JavaScript Functionality

■ Dynamic HTML for Microsoft TV

■ Summary of the Microsoft TV Object Model

■ JavaScript 1.1 References

The specification adopted by the Advanced Television Enhancement Forum (ATVEF) defines Level 0 Document Object Model (Level 0 DOM) and ECMA 262 Language specification (ECMAScript) as the standards for scripting interactive TV Web pages. There is no official Level 0 DOM specification or recommendation. The World Wide Web Consortium (W3C) provides a recommendation for Level 1 DOM only, which states that Level 1 DOM "...is a platform and language-neutral interface that will allow programs and scripts to dynamically access and update the content, structure, and style of documents." Fortunately, the specification produced by the ATVEF wisely notes that Level 0 DOM and ECMAScript are equivalent to JavaScript 1.1. Therefore, this book follows the ATVEF lead and uses the JavaScript 1.1 object model as the basis for scripting for Microsoft TV.

Using JavaScript 1.1 as the baseline for scripting interactive TV pages is both liberating and restrictive. It is liberating because having a solid, well-defined object model takes the guesswork out of trying to script pages based on several different, and sometimes conflicting, object models. On the other hand, using JavaScript 1.1 can feel restrictive because, at the time of this writing, JavaScript 1.2 is both the most current version of JavaScript and the version of JavaScript used in conjunction with dynamic HTML (DHTML). However, JavaScript 1.1 is often the version industrial-based sites use to avoid browser incompatibilities relating to the different object models supported by Netscape Navigator 4.0 and Microsoft Internet Explorer 4.0 and later.

We have provided the following table showing browser versions and the JavaScript that they support to give you an idea of JavaScript 1.1's historical place in browser evolution.

Browser Version	Netscape Navigator	Microsoft Internet Explorer
2	JavaScript 1.0	
3	JavaScript 1.1	JScript 1.0
4	JavaScript 1.2; not fully ECMA-262 compliant	JavaScript 1.2; fully ECMA-262 compliant

JavaScript 1.1, generally considered the industrial scripting standard, may not offer all the features of DHTML, but it does offer several important advantages:

■   It complies with content creation standards adopted by ATVEF, which ensures that scripts will run reliably across a wide variety of TV receivers.

■   It works across multiple Web browsers without additional code to handle object model differences.

THE "MICROSOFT TV PROGRAMMER'S GUIDE"

The "Document Object Model" topic in the "Microsoft TV Programmer's Guide" on the companion CD contains complete documentation on the objects, properties, methods, and events supported by Microsoft TV. The Microsoft TV object model, based on JavaScript 1.1, enables referencing and manipulating of the main elements on a Web page, including windows, documents, frames, images, buttons, forms, and form components (such as list boxes, check boxes, and radio buttons). The object model is shown in Figure 24.1.

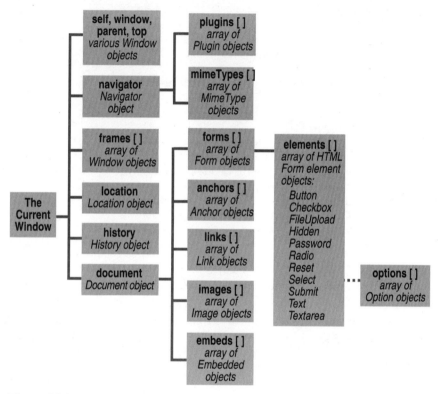

Figure 24.1 *The Microsoft TV object model.*

USING JAVASCRIPT WITH MICROSOFT TV

Microsoft TV provides robust support for JavaScript 1.1. However, there are some limitations, which you can deal with using the methods shown in the following text.

Referencing Objects with JavaScript

When referencing Microsoft TV objects using JavaScript, it is best to reference objects by *name* or by *index*, rather than *id* as is done with DHTML. To illustrate this point, let's take a look at how you can reference and manipulate the frame object in Microsoft TV using JavaScript 1.1. First, look at the frameset that builds the page. The name of the frame that we will reference is "content" (shown in bold in the following code):

```
<html>
<head>
<title>exotic excursions</title>
</head>
<script type="text/JavaScript">
if (navigator.appversion.indexof("webtv") != -1 )
{
    document.write("<body hspace=0 vspace=0>");
}
</script>

<frameset cols="209,*" framespacing=0 frameborder=0 hspace=0 vspace=0
marginwidth=0 marginheight=0>
<frame name="tv" src="exotic_tv.html" marginwidth=0 marginheight=0
scrolling=no noresize frameborder=no>
<frame name="content" src="exotic_feature1.html" marginwidth=0
marginheight=0 scrolling=no noresize frameborder=no>
</frameset>
</html>
```

Next, look at sample JavaScript on the exotic_tv.html page in the "tv" frame that loads a new page into the content frame. As you can see in the featureload( ) function (shown in bold), the frame is referenced as parent.frames["content"].location.href, where *parent* is the main window, *frames* is the frame array, and *content* is the frame within the frames array.

```
<html>
<head>
<script type="text/JavaScript">

function featureload()
{
    parent.frames["content"].location.href="exotic_feature1.html";
}

</script>
<title>exotic excursions</title>
</head>
<body bgcolor=#000000 leftmargin=0 topmargin=0>

<div style="position:absolute;top:243;left:16">
```

```
<a href="javascript:onclick=featureload()"><img src="images/feature2.gif"
width=190 height=42 alt="" border=0 name="btn_feature"></a>
</div>

</body>
</html>
```

Enabling the *onClick* Event with the Image Object

With JavaScript 1.1, the image object does not support the *onClick* event. However, an easy workaround to enable the *onClick* event for an image is to surround the image with an anchor and then call a JavaScript function from the anchor tag.

For more information about enabling the *onClick* event for the image object, see the "Making an Image Clickable" section in Chapter 11, "Adding Images to Microsoft TV Content."

UNIMPLEMENTED JAVASCRIPT FUNCTIONALITY

Microsoft TV 1.0 does not implement the following JavaScript functionality:

- **Multiple windows**—Because Microsoft TV does not support multiple windows, scripts cannot open windows and subsequently read from, or write to, those windows. An exception is made, however, for windows larger than 400-by-300 pixels, which are treated as a new page. The user can get back to the original window by clicking the Back button. The lack of multiple windows has caused problems for authors who use the window object to refer to frames. For instance:

  ```
  window.open("ft-top.html","top");
  window.open("ft-left.html","left");
  ```

  would not work on Microsoft TV. Instead, consider using:

  ```
  parent.frames[0].location = 'ft-top.html';
  parent.frames[1].location = 'ft-left.html';
  ```

- **Regular expressions**—For example: var re = /a|b|c/i is not supported.

- **Signed scripts**—Scripts digitally signed for security are not supported.

- **Adding values to built-in objects**—For example, document.myVar.color is not supported; use myVar.color instead.

- **Numeric sort**—The array.sort method in JavaScript does not support the optional function parameter to customize the sort order. To do a numeric sort, use a simple "bubble sort" function such as the one following. Please note that this type of sort may not be efficient for large arrays.

```
function bubbleSort(arrayName,arrayLength)
{
    for (var i=0; i<(length-1); i++)
        for (var j=i+1; j<length; j++)
            if (arrayName[j] < arrayName[i])
            {
                var dummy = arrayName[i];
                arrayName[i] = arrayName[j];
                arrayName[j] = dummy;
            }
}
```

Microsoft TV implements the following items differently than you may expect:

■ **Arrays**—Microsoft TV limits arrays to 32,768 elements.

■ **Errors**—Microsoft TV does not alert users when a Web page contains JavaScript errors, although serious errors will disable JavaScript on a given page.

■ **Math**—Microsoft TV uses 32-bit floats instead of 64-bit doubles. This is adequate for all but high-end mathematical computations.

Unimplemented Functions and Event Handlers

Microsoft TV 1.0 does not implement certain JavaScript functions, including:

```
object.prototype(), document.domain, form.encoding, history.next,
window.onError(), window.onUnload()
```

■ window.opener (Microsoft TV supports read-only for window.opener rather than read/write.)

The following event handlers are not supported:

■ onAbort, onError, onUnload, keyDown, keyUp, keyPress, mouseDown, mouseMove

DYNAMIC HTML FOR MICROSOFT TV

Unlike DHTML, JavaScript 1.1 does not allow the referencing or manipulating of each element on the page as an individual object. However, Microsoft TV supports a limited subset of DHTML that enables you to reference individual objects; to hide, show, and move divisions; and to dynamically format elements on a page. For more information about Microsoft TV's DHTML support, see Chapter 25, "DHTML for Microsoft TV."

SUMMARY OF THE MICROSOFT TV OBJECT MODEL

Here is a quick reference to the objects, properties, methods, and events supported by Microsoft TV 1.0. For more detailed information about the object model, see the "Object Model for Microsoft TV" heading in the "Microsoft TV Programmer's Guide" section of the companion CD.

Object	Properties	Methods	Events
Anchor	eval toString valueOf	eval	none
Anchors Array	length toString valueOf	eval	none
Area	hash host hostname href pathname port protocol search target	eval toString valueOf	onMouseOut onMouseOver
Array	length prototype	eval join reverse sort toString valueOf	none
Arguments Array	length	eval toString valueOf	none
Button	form name type value	blur click eval focus toString valueOf	onBlur onClick onFocus

(continued)

continued

Object	Properties	Methods	Events
Checkbox	checked	blur	onBlur
	defaultChecked	click	onClick
	form	eval	onFocus
	name	focus	
	type	toString	
	value	valueOf	
Date	prototype	eval	none
	getDate		
	getDay		
	getHours		
	getMinutes		
	getMonth		
	getSeconds		
	getTime		
	getTimezoneOffset		
	getYear		
	parse		
	setDate		
	setHours		
	setMinutes		
	setMonth		
	setSeconds		
	setTime		
	setYear		
	toGMTString		
	toLocaleString		
	toString		
	UTC		
	valueOf		
Document	alinkColor	close	none
	Anchor	eval	
	anchors	open	
	Area	toString	
	bgColor	UTC	
	cookie	valueOf	
	domain	write	
	embeds	writeln	
	fgColor		
	Form		
	forms		
	Image		
	images		
	lastModified		

Object	Properties	Methods	Events
	linkColor		
	Link		
	links		
	referrer		
	title		
	URL		
	vlinkColor		
Elements Array	length	eval	none
		toString	
		valueOf	
Embeds Array	length	eval	none
	toString		
	valueOf		
FileUpload	form	blur	onBlur
	name	eval	onChange
	type	focus	onFocus
	value	toString	
		valueOf	
Form	action	blur	onBlur
	Button	focus	onChange
	Checkbox	eval	onFocus
	elements	toString	
	encoding	valueOf	
	FileUpload		
	Hidden		
	Length		
	method		
	name		
	Password		
	Radio		
	Reset		
	Select		
	Submit		
	target		
	Text		
	Textarea		
Forms Array	length	eval	none
	toString		
	valueOf		

(continued)

continued

Object	Properties	Methods	Events
Frame	frames name length parent self window	blur clearTimeout eval focus toString valueOf	onBlur onFocus
Frames Array	length	eval toString valueOf	none
Function	length toString valueOf	eval	none
Hidden	name type value	eval toString valueOf	none
History	current length next previous	back eval forward go toString valueOf	none
History Array	length toString valueOf	eval	none
Image	border complete height hspace lowsrc name prototype src vspace width	eval toString valueOf	onAbort (not supported by Microsoft TV) onError (not supported by Microsoft TV) onLoad
Images Array	length	eval toString valueOf	none
Link and Area	hash host hostname href pathname	eval toString valueOf	onClick onMouseOut onMouseOver

Object	Properties	Methods	Events
	port protocol search target		
Links Array	length toString valueOf	eval	none
Location	hash host hostname href pathname port protocol search	eval reload replace toString valueOf	none
Math	E LN LN10 LOG2E LOG10E PI SQRT1_1 SQRT2	abs acos asin atan atan2 ceil cos exp floor log max min pow random round sin sqrt tan toString valueOf	none
MimeType	description enabledPlugin type suffixes	eval toString valueOf	none
MimeTypes Array	length toString valueOf	eval	none

(continued)

continued

Object	Properties	Methods	Events
Navigator	appCodeName appName appVersion mimeTypes plugins userAgent	eval javaEnabled taintEnabled toString valueOf	none
Number	MAX_VALUE MIN_VALUE NaNvalueOf NEGATIVE_INFINITY POSITIVE_INFINITY prototype	eval toString valueOf	none
Object	constructor	eval toString valueOf	none
Options Array	length toString valueOf	eval	none
Options Array Elements	defaultSelected index length selected selectedIndex text value	eval toString valueOf	none
Password	defaultValue form name type value text value	eval focus select toString valueOf	onFocus
Plugin	description filename length name	eval toString valueOf	none
Plugins Array	length refresh toString valueOf	eval	none

Object	Properties	Methods	Events
Radio	checked defaultChecked form length name type value	blur click eval focus toString valueOf	none
Reset	form name type value	blur click eval focus toString valueOf	onBlur onClick onFocus
Select	form length name options selectedIndex text type	blur eval focus toString valueOf	onBlur onClick onFocus
String	length prototype	anchor big blink bold charAt eval fixed fontcolor fontsize indexOf italics lastIndexOf link small split strike sub substring sup toLowerCase toString toUpperCase valueOf	none

(continued)

continued

Object	Properties	Methods	Events
Submit	eval form name toString type value valueOf	blur click focus	onBlur onClick onFocus·
Text	defaultValue form name type value	blur eval focus toString select valueOf	onBlur onChange onClick onFocus
Textarea	defaultValue form name type value	blur eval focus toString select valueOf	onBlur onChange onClick onFocus
Window	closed defaultStatus document Frame frames history length location name opener parent self status top window	alert blur clearTimeout close confirm focus open prompt setTimeOut toString valueOf	onBlur onError (not supported by Microsoft TV) onFocus onLoad onUnload (not supported by Microsoft TV)

JavaScript 1.1 References

For more information about JavaScript 1.1, see Netscape's JavaScript 1.1 Guide at: *http://home.netscape.com/eng/mozilla/3.0/handbook/javascript/index.html.*

WHAT'S NEXT

This chapter briefly covered the document object model supported by Microsoft TV, described how to reference objects in Microsoft TV, and then listed JavaScript functionality not supported in Microsoft TV. This chapter also provided a quick summary of the objects, properties, events, and methods supported by Microsoft TV. The next chapter takes a look at the subset of DHTML elements and properties supported by Microsoft TV.

DHTML for Microsoft TV

In This Chapter

■ DHTML Elements and Properties Supported by Microsoft TV

■ Using the "Microsoft TV Programmer's Guide"

■ A Demonstration of DHTML in Action

Microsoft TV supports a subset of dynamic HTML (DHTML)—HTML 4.0, Cascading Style Sheets (CSS) level 1 and level 2, JavaScript, and a Document Object Model (DOM)—that enables all HTML elements on a Web page to be referenced as individual objects. With the supported DHTML subset, content developers can dynamically hide and show divisions, move divisions, reformat styles, and determine the height and width of elements. Developers should use DHTML cautiously, however, when designing interactive TV content, for the following reasons:

■ For Microsoft TV, DHTML can only be used reliably with <div> tags, because only <div> tags support the class, id, and style attributes.

■ DHTML has not been adopted by the Advanced Television Enhancement Forum (ATVEF), so content created using DHTML may not function properly on all interactive TV receivers.

■   DHTML is not an industry-wide standard. In fact, creating Web pages that reliably run DHTML on Microsoft Internet Explorer and Netscape Navigator is a challenging task, because Internet Explorer 4.0 and later and Netscape Navigator 4.0 and later have different object models.

■   At first glance, it may seem like a good idea to create interactive TV content with one page that consists of the video object (either full-screen or reduced size) and several divisions that are dynamically hidden or displayed. This seems particularly appealing for layering content over full-screen video. However, this solution breaks down when you attempt to read or write to a server database because these actions cause the page hosting the video control to refresh, resulting in a flicker or interruption in the TV picture. For this reason, a frames-based solution, in which the video object is hosted on a static page in a frame, is used in most of the *Building Interactive Entertainment and E-Commerce Content for Microsoft TV* samples.

The combination of HTML 4.0, CSS1 and CSS2, and JavaScript 1.1 can meet the majority of a developer's needs, but developers may occasionally use DHTML to perform certain tasks. This chapter provides guidelines for using DHTML for Microsoft TV, and it lists the subset of DHTML objects and properties that are supported.

DHTML ELEMENTS AND PROPERTIES SUPPORTED BY MICROSOFT TV

Microsoft TV provides support for elements and properties through the use of *document.all[ ]*, a versatile array that contains all the HTML elements in a document. Using *document.all[ ]*, you can reference any element in a document and then read or modify the subset of supported properties for the element, as listed in the following table.

Property	*Description*
document.all	An array of all HTML elements in a document.
id	The value of the id attribute.
offsetHeight	The height of the element.
offsetLeft	The X coordinate of the element.
offsetTop	The Y coordinate of the element.
offsetWidth	The width of the element.
disabled	Indicates whether an element is available for user interaction.

Property	Description
readOnly	Indicates whether an element can be edited on the page by the TV viewer.
style	The inline cascading style sheets' style attributes for an element. For more information about supported style attributes, see "CSS for Microsoft TV" in the "Microsoft TV Programmer's Guide" on the *Building Interactive Entertainment and E-Commerce Content for Microsoft TV* companion CD.
visibility	Indicates whether the element is visible on the page.

USING THE "MICROSOFT TV PROGRAMMER'S GUIDE"

The "DHTML for Microsoft TV" topic in the "Microsoft TV Programmer's Guide" on the companion CD provides complete documentation of the supported subset of DHTML functionality.

A DEMONSTRATION OF DHTML IN ACTION

The following example demonstrates how DHTML can be applied to dynamically alter a document. It also illustrates the syntax used to reference an HTML element and to change its properties. This example determines whether the description text for an anchor is hidden or visible when the anchor receives the focus, as indicated by the yellow highlight box for Microsoft TV. Also, the Enabled and Disabled radio buttons set the state of the disabled property of the "Disabled Test" checkbox.

```
<script language="JavaScript">
function showone()
{
    document.all.one_description.style.visibility = "visible"
}
function hideone()
{
    document.all.one_description.style.visibility = "hidden"
}
function showtwo()
{
    document.all.two_description.style.visibility = "visible"
}
function hidetwo()
{
    document.all.two_description.style.visibility = "hidden"
}
```

(continued)

```
function enable()
{
    document.all.checkbox1.disabled = false
}
function disable()
{
    document.all.checkbox1.disabled = true
}

</script>

<html>
<head>
<title>dhtml example</title>
</head>
<body>
<div style="position:absolute;top:25">
<a href="pghtmlsamp2.html" onmouseover="showone()"
onmouseout="hideone()">option one</a><br><br>

<a href ="pghtmlsamp2.html" onmouseover="showtwo()"
onmouseout="hidetwo()">option two</a>
</div>

<div style="position:absolute;top:0;visibility:hidden"
id="one_description">option one description</div>

<div style="position:absolute;top:0;visibility:hidden"
id="two_description">option two description</div>

<div style="position:absolute;top:120" id="form1">
<form action="none" method="">
<input type="radio" name="optionbutton1" value="enabled" checked
onclick="enable()">enabled<br>
<input type="radio" name="optionbutton1" value="enabled"
onclick="disable()">disabled<br><br>
<input type="checkbox" id="checkbox1" value="test">disabled test
</form>
</div>
</body>
</html>
```

WHAT'S NEXT

This chapter offered a quick glimpse at Microsoft TV's DHTML support. The following chapter takes a look at Microsoft TV's support for ECMAScript.

Chapter 26

ECMAScript for Interactive TV

In This Chapter

■   Determining the Microsoft TV Object Model

■   Specifying ECMAScript as the Scripting Language

■   Embedding ECMAScript in HTML

■   Determining Which ECMAScript Statements Are Supported by Level 0 DOM

■   ECMAScript and JavaScript References

The specification adopted by the Advanced Television Enhancement Forum (ATVEF) defines the ECMA 262 Language specification (ECMAScript) as the standard for scripting interactive TV content. ECMAScript is a standardized version of JavaScript that has been created by the European Computer Manufacturers Association (ECMA). ECMAScript, based on Netscape's JavaScript and Microsoft's offering of the language called JScript, creates a standardized version of JavaScript (a name owned by Netscape) that favors neither JavaScript nor JScript. Not all implementations of JavaScript currently conform to all the details of the ECMA 262 Standard.

> **NOTE** Because the ECMA 262 Standard is complete and freely accessible, it is not included on our *Building Interactive Entertainment and E-Commerce Content for Microsoft TV* companion CD. View the ECMA Standard atm *http://www.ecma.ch/stand/ecma-262.htm.*

DETERMINING THE MICROSOFT TV OBJECT MODEL

The specification developed by the ATVEF defines ECMAScript as the scripting language standard and it defines Level 0 Document Object Model (Level 0 DOM) as the object model standard. However, there is no formal specification available for the Level 0 DOM. As previously mentioned in Chapter 24, "Document Object Model for Microsoft TV," the World Wide Web Consortium (W3C) provides a recommendation for Level 1 DOM, but not for Level 0 DOM. However, the ATVEF resolves the lack of an official Level 0 DOM by stating that ECMAScript and Level 0 DOM are equivalent to JavaScript 1.1. As a result, the Microsoft TV 1.0 object model is based on JavaScript 1.1. See "Object Model for Microsoft TV" in the "Microsoft TV Programmer's Guide" on the companion CD.

SPECIFYING ECMASCRIPT AS THE SCRIPTING LANGUAGE

Client-side ECMAScript scripts are part of an HTML file and are usually coded within <script> </script> tags. Traditionally, a Web page specifies its scripting language by naming the language in the <script> tag. For example: <script language>"JavaScript">.

However, rather than using the *language* property, the preferred ATVEF method for interactive TV is to use the *type* property as follows:

```
<script type="text/JavaScript">
```

The ATVEF prefers this syntax because it refers to an open MIME type that is not "owned" by anyone.

EMBEDDING ECMASCRIPT IN HTML

There are several ways to embed ECMAScript into an HTML page:

- Between a pair of <script> </script> tags

- Using an event handler, such as *onClick* or *onMouseOver*, in an HTML tag

- As the body of a URL using the syntax *JavaScript: statement*

Each of these methods is discussed in the following sections.

Embedding ECMAScript Between <script> </script> Tags

For some pages, it is useful to execute ECMAScript when the page loads. To do this, place the <script> </script> tags at the top of the HTML document as shown in the following example:

```
<html>
<head>
<script type="text/JavaScript">
document.write('<h1>this is written with ECMAScript code.</h1>')
</script>
<title>untitled</title>
</head>


<body>
This text is written in the body area.
</body>
</html>
```

Embedding ECMAScript Using an Event Handler

ECMAScript, like JavaScript, provides a series of event handlers for certain HTML tags that enable you to specify script to execute when a particular event, such as *onClick*, *onMouseOver*, or *onLoad*, is fired. The most common way to implement an event handler is to create a function within <script> </script> tags and then invoke the function from the event handler. The following example shows an event handler specified for the Go button. When a user clicks the Go button, the *onClick* event is fired, and the color_go function is called and executed. For this example, the color_go function navigates to the page selected in the color choice list box.

```
<html>
<script type="text/JavaScript">
function color_go()
{
    if (document.color_choice.go.selectedindex == 0)
    {
        window.location = "blue.html";
    }
    else if (document.color_choice.go.selectedindex == 1)
    {
        window.location = "red.html";
    }
}
</script>
```

(continued)

```
<head>
<title>untitled</title>
</head>

<body>

<form name="color_choice" action="" method="">
<select name="go">
<option value="one">blue
<option value="two">red
</select>

<input type="button" name="go" value="go" onclick="color_go()">
</form>
</body>
</html>
```

Embedding ECMAScript in a URL

The *JavaScript: statement* can be used anywhere a URL is used. This method is a good one to use when executing event handlers for the <img> tag, because the <img> tag does not support the *onClick* event handler. To load a page, for example, a user can evade this limitation by making the image function as a link and then specifying *JavaScript* instead of a URL.

The following code shows a simplified version of the ECMAScript code (used in other samples in this book) that returns the view to a full-screen TV page from an interactive TV page. In the example, an image is made to function as a link, and then the JavaScript:onclick=sample() is specified as the URL. This calls the sample function when the user clicks the image.

```
<html>
<head>
<script type="text/JavaScript">
function sample()
{
    alert('You can embed JavaScript in an anchor tag.')
}
</script>

<title>sample snippet</title>
</head>
<body>
<a href="javascript:onclick=sample()"><img src="images/sample.gif"
width=102 height=20 border="0"></a>
</body>
</html>
```

DETERMINING WHICH ECMASCRIPT STATEMENTS ARE SUPPORTED BY LEVEL 0 DOM

The Microsoft TV object model on the companion CD provides a list of the objects, properties, methods, and events that are supported by Level 0 DOM. But it does not state which ECMAScript statements are supported; you can find that information in the ECMA 262 Standard. This standard can make for some pretty dry reading, so a time-saving list of statements supported by the ECMA Standard is printed below.

Statement	*Syntax*	*Purpose*
Block	{statement list}	Combines a number of statements into a statement block
Break	break; break label;	Exits from the innermost loop or from the statement named by label
Continue	continue continue label;	Restarts the innermost loop or the loop named by label
Empty	;	Does nothing
For	For (initialize; test; increment) *statement*	An easy-to-use test
for/in	For (variable in object) *statement*	Loops through properties of an object

Function	Function *funcname* ([arg1 [...,argn]]) {
	statements
	}
		Declares a function

if/else	if (expression) *statement1* *[else statement2]*	Conditionally executes code
return	Return [expression] ;	Returns from a function or returns the value of expression from a function
var	Var name1 *[=value_1]* *[...,name_n [=value_n]];*	Declares and initializes variables
with	with *(object)* *statement*	Extends the current scope chain
while	while *(expression)* *statement*	A basic loop construct

ECMASCRIPT AND JAVASCRIPT REFERENCES

ECMAScript is thoroughly documented in the ECMA 262 Standard, but the documentation does not provide practical examples or code snippets that you can use in your interactive TV content. Therefore, to create ECMAScript for interactive TV pages, we suggest reviewing the following documentation:

- **ECMA 262 Standard**—The ECMA 262 Standard can be found at *http://www.ecma.ch/stand/ecma-262.htm.*

- **The Object Model for Microsoft TV**—Information on this topic can be found in the *Building Interactive Entertainment and E-Commerce Content for Microsoft TV* companion CD.

- **Netscape's JavaScript 1.1 Documentation**—This document can be found on Netscape's Web site at *http://home.netscape.com/eng/mozilla/3.0/handbook/javascript/index.html.*

Index

Note: Page numbers in italics refer to figures, tables, or illustrations.

Index

The manuscript for this book was prepared using Microsoft Word 2000. Pages were composed by Microsoft Press using Adobe PageMaker 6.52 for Windows, with text in Garamond and display type in Helvetica Black. Composed pages were delivered to the printer as electronic prepress files.

Cover Graphic Designer
Tom Draper Design

Cover Illustrator
Tom Draper Design

Interior Graphic Artists
Joel Panchot and Julie Hammerquist

Principal Compositor
Carl Diltz

Principal Proofreader/Copy Editor
Holly Viola

Indexer
Liz Cunningham

SYSTEM REQUIREMENTS

The companion CD-ROM for *Building Interactive Entertainment and E-Commerce Content for Microsoft TV* requires the following hardware and software to ensure that its contents display properly and that all the features work:

- Pentium processor running at 200 MHz or faster

- 32 MB RAM

- 6x or higher CD-ROM drive

- 16-bit or higher sound card and speakers

- 16-bit or higher video card

- 800 x 600 or higher resolution

- Microsoft Windows 95, Windows 98, Windows NT 4.0 (with Service Pack 3 or later installed), or Windows 2000 Professional

- Internet Explorer 4.01 or later

MICROSOFT LICENSE AGREEMENT

Book Companion CD

SOFTWARE PRODUCT LICENSE

The SOFTWARE PRODUCT is protected by United States copyright laws and international copyright treaties, as well as other intellectual property laws and treaties. The SOFTWARE PRODUCT is licensed, not sold.

1. GRANT OF LICENSE. This EULA grants you the following rights:

    a. Software Product. You may install and use one copy of the SOFTWARE PRODUCT on a single computer. The primary user of the computer on which the SOFTWARE PRODUCT is installed may make a second copy for his or her exclusive use on a portable computer.

    b. Storage/Network Use. You may also store or install a copy of the SOFTWARE PRODUCT on a storage device, such as a network server, used only to install or run the SOFTWARE PRODUCT on your other computers over an internal network; however, you must acquire and dedicate a license for each separate computer on which the SOFTWARE PRODUCT is installed or run from the storage device. A license for the SOFTWARE PRODUCT may not be shared or used concurrently on different computers.

    c. License Pak. If you have acquired this EULA in a Microsoft License Pak, you may make the number of additional copies of the computer software portion of the SOFTWARE PRODUCT authorized on the printed copy of this EULA, and you may use each copy in the manner specified above. You are also entitled to make a corresponding number of secondary copies for portable computer use as specified above.

    d. Sample Code. Solely with respect to portions, if any, of the SOFTWARE PRODUCT that are identified within the SOFTWARE PRODUCT as sample code (the "SAMPLE CODE"):

        i. Use and Modification. Microsoft grants you the right to use and modify the source code version of the SAMPLE CODE, *provided* you comply with subsection (d)(iii) below. You may not distribute the SAMPLE CODE, or any modified version of the SAMPLE CODE, in source code form.

        ii. Redistributable Files. Provided you comply with subsection (d)(iii) below, Microsoft grants you a nonexclusive, royalty-free right to reproduce and distribute the object code version of the SAMPLE CODE and of any modified SAMPLE CODE, other than SAMPLE CODE (or any modified version thereof) designated as not redistributable in the Readme file that forms a part of the SOFTWARE PRODUCT (the "Non-Redistributable Sample Code"). All SAMPLE CODE other than the Non-Redistributable Sample Code is collectively referred to as the "REDISTRIBUTABLES."

        iii. Redistribution Requirements. If you redistribute the REDISTRIBUTABLES, you agree to: (i) distribute the REDISTRIBUTABLES in object code form only in conjunction with and as a part of your software application product; (ii) not use Microsoft's name, logo, or trademarks to market your software application product; (iii) include a valid copyright notice on your software application product; (iv) indemnify, hold harmless, and defend Microsoft from and against any claims or lawsuits, including attorney's fees, that arise or result from the use or distribution of your software application product; and (v) not permit further distribution of the REDISTRIBUTABLES by your end user. Contact Microsoft for the applicable royalties due and other licensing terms for all other uses and/or distribution of the REDISTRIBUTABLES.

2. DESCRIPTION OF OTHER RIGHTS AND LIMITATIONS.

    • **Limitations on Reverse Engineering, Decompilation, and Disassembly.** You may not reverse engineer, decompile, or disassemble the SOFTWARE PRODUCT, except and only to the extent that such activity is expressly permitted by applicable law notwithstanding this limitation.

    • **Separation of Components.** The SOFTWARE PRODUCT is licensed as a single product. Its component parts may not be separated for use on more than one computer.

    • **Rental.** You may not rent, lease, or lend the SOFTWARE PRODUCT.

    • **Support Services.** Microsoft may, but is not obligated to, provide you with support services related to the SOFTWARE PRODUCT ("Support Services"). Use of Support Services is governed by the Microsoft policies and programs described in the user manual, in "on-line" documentation, and/or in other Microsoft-provided materials. Any supplemental software code provided to you as part of the Support Services shall be considered part of the SOFTWARE PRODUCT and subject to the terms and conditions of this EULA. With respect to technical information you provide to Microsoft as part of the Support Services, Microsoft may use such information for its business purposes, including for product support and development. Microsoft will not utilize such technical information in a form that personally identifies you.

- **Software Transfer.** You may permanently transfer all of your rights under this EULA, provided you retain no copies, you transfer all of the SOFTWARE PRODUCT (including all component parts, the media and printed materials, any upgrades, this EULA, and, if applicable, the Certificate of Authenticity), **and** the recipient agrees to the terms of this EULA.

- **Termination.** Without prejudice to any other rights, Microsoft may terminate this EULA if you fail to comply with the terms and conditions of this EULA. In such event, you must destroy all copies of the SOFTWARE PRODUCT and all of its component parts.

3. **COPYRIGHT.** All title and copyrights in and to the SOFTWARE PRODUCT (including but not limited to any images, photographs, animations, video, audio, music, text, SAMPLE CODE, REDISTRIBUTABLES, and "applets" incorporated into the SOFTWARE PRODUCT) and any copies of the SOFTWARE PRODUCT are owned by Microsoft or its suppliers. The SOFTWARE PRODUCT is protected by copyright laws and international treaty provisions. Therefore, you must treat the SOFTWARE PRODUCT like any other copyrighted material **except** that you may install the SOFTWARE PRODUCT on a single computer provided you keep the original solely for backup or archival purposes. You may not copy the printed materials accompanying the SOFTWARE PRODUCT.

4. **U.S. GOVERNMENT RESTRICTED RIGHTS.** The SOFTWARE PRODUCT and documentation are provided with RESTRICTED RIGHTS. Use, duplication, or disclosure by the Government is subject to restrictions as set forth in subparagraph (c)(1)(ii) of the Rights in Technical Data and Computer Software clause at DFARS 252.227-7013 or subparagraphs (c)(1) and (2) of the Commercial Computer Software—Restricted Rights at 48 CFR 52.227-19, as applicable. Manufacturer is Microsoft Corporation/One Microsoft Way/Redmond, WA 98052-6399.

5. **EXPORT RESTRICTIONS.** You agree that you will not export or re-export the SOFTWARE PRODUCT, any part thereof, or any process or service that is the direct product of the SOFTWARE PRODUCT (the foregoing collectively referred to as the "Restricted Components"), to any country, person, entity, or end user subject to U.S. export restrictions. You specifically agree not to export or re-export any of the Restricted Components (i) to any country to which the U.S. has embargoed or restricted the export of goods or services, which currently include, but are not necessarily limited to, Cuba, Iran, Iraq, Libya, North Korea, Sudan, and Syria, or to any national of any such country, wherever located, who intends to transmit or transport the Restricted Components back to such country; (ii) to any end user who you know or have reason to know will utilize the Restricted Components in the design, development, or production of nuclear, chemical, or biological weapons; or (iii) to any end user who has been prohibited from participating in U.S. export transactions by any federal agency of the U.S. government. You warrant and represent that neither the BXA nor any other U.S. federal agency has suspended, revoked, or denied your export privileges.

6. **NOTE ON JAVA SUPPORT.** THE SOFTWARE PRODUCT MAY CONTAIN SUPPORT FOR PROGRAMS WRITTEN IN JAVA. JAVA TECHNOLOGY IS NOT FAULT TOLERANT AND IS NOT DESIGNED, MANUFACTURED, OR INTENDED FOR USE OR RESALE AS ON-LINE CONTROL EQUIPMENT IN HAZARDOUS ENVIRONMENTS REQUIRING FAIL-SAFE PERFORMANCE, SUCH AS IN THE OPERATION OF NUCLEAR FACILITIES, AIRCRAFT NAVIGATION OR COMMUNICATION SYSTEMS, AIR TRAFFIC CONTROL, DIRECT LIFE SUPPORT MACHINES, OR WEAPONS SYSTEMS, IN WHICH THE FAILURE OF JAVA TECHNOLOGY COULD LEAD DIRECTLY TO DEATH, PERSONAL INJURY, OR SEVERE PHYSICAL OR ENVIRONMENTAL DAMAGE. SUN MICROSYSTEMS, INC. HAS CONTRACTUALLY OBLIGATED MICROSOFT TO MAKE THIS DISCLAIMER.

DISCLAIMER OF WARRANTY

NO WARRANTIES OR CONDITIONS. MICROSOFT EXPRESSLY DISCLAIMS ANY WARRANTY OR CONDITION FOR THE SOFTWARE PRODUCT. THE SOFTWARE PRODUCT AND ANY RELATED DOCUMENTATION ARE PROVIDED "AS IS" WITHOUT WARRANTY OR CONDITION OF ANY KIND, EITHER EXPRESS OR IMPLIED, INCLUDING, WITHOUT LIMITATION, THE IMPLIED WARRANTIES OF MERCHANTABILITY, FITNESS FOR A PARTICULAR PURPOSE, OR NONINFRINGEMENT. THE ENTIRE RISK ARISING OUT OF USE OR PERFORMANCE OF THE SOFTWARE PRODUCT REMAINS WITH YOU.

LIMITATION OF LIABILITY. TO THE MAXIMUM EXTENT PERMITTED BY APPLICABLE LAW, IN NO EVENT SHALL MICROSOFT OR ITS SUPPLIERS BE LIABLE FOR ANY SPECIAL, INCIDENTAL, INDIRECT, OR CONSEQUENTIAL DAMAGES WHATSOEVER (INCLUDING, WITHOUT LIMITATION, DAMAGES FOR LOSS OF BUSINESS PROFITS, BUSINESS INTERRUPTION, LOSS OF BUSINESS INFORMATION, OR ANY OTHER PECUNIARY LOSS) ARISING OUT OF THE USE OF OR INABILITY TO USE THE SOFTWARE PRODUCT OR THE PROVISION OF OR FAILURE TO PROVIDE SUPPORT SERVICES, EVEN IF MICROSOFT HAS BEEN ADVISED OF THE POSSIBILITY OF SUCH DAMAGES. IN ANY CASE, MICROSOFT'S ENTIRE LIABILITY UNDER ANY PROVISION OF THIS EULA SHALL BE LIMITED TO THE GREATER OF THE AMOUNT ACTUALLY PAID BY YOU FOR THE SOFTWARE PRODUCT OR US$5.00; PROVIDED, HOWEVER, IF YOU HAVE ENTERED INTO A MICROSOFT SUPPORT SERVICES AGREEMENT, MICROSOFT'S ENTIRE LIABILITY REGARDING SUPPORT SERVICES SHALL BE GOVERNED BY THE TERMS OF THAT AGREEMENT. BECAUSE SOME STATES AND JURISDICTIONS DO NOT ALLOW THE EXCLUSION OR LIMITATION OF LIABILITY, THE ABOVE LIMITATION MAY NOT APPLY TO YOU.

MISCELLANEOUS

This EULA is governed by the laws of the State of Washington USA, except and only to the extent that applicable law mandates governing law of a different jurisdiction.

Should you have any questions concerning this EULA, or if you desire to contact Microsoft for any reason, please contact the Microsoft subsidiary serving your country, or write: Microsoft Sales Information Center/One Microsoft Way/Redmond, WA 98052-6399.

For information about Microsoft Press®
products, visit our Web site at
mspress.microsoft.com